HENRY CORT: THE GREAT FINER
Creator of Puddled Iron

Bas-relief of Henry Cort in Hampstead Church

HENRY CORT: THE GREAT FINER

Creator of Puddled Iron

by R. A. Mott
edited by Peter Singer

The Metals Society
London
1983

Book 299
published in 1983
in association with The Historical Metallurgy Society
by The Metals Society
1 Carlton House Terrace
London SW1Y 5DB

© The Metals Society 1983

All rights reserved
ISBN 0 904357 55 4

Printed and made in England by
Adlard and Son Ltd, Dorking, Surrey

CONTENTS

Introduction by the Editor.. ix

Preface by R. A. Mott... xiii

CHAPTER ONE
The search for new finery processes............................... 1

CHAPTER TWO
The background of Henry Cort: Navy Agent.................. 16

CHAPTER THREE
The early history of Fareham Ironworks:
Fontley and Gosport.. 22

CHAPTER FOUR
The development of puddling and rolling
at Fareham.. 27

CHAPTER FIVE
The rolling of metals... 31

CHAPTER SIX
Henry Cort's puddling process....................................... 37

CHAPTER SEVEN
Testing Cort's iron in the Naval Dockyards
1783–6 40

CHAPTER EIGHT
The adoption of Cort's processes
(1) Coalbrookdale 47

CHAPTER NINE
The adoption of Cort's processes
(2) Developments at Cyfarthfa 1787–9 51

v

Contents

CHAPTER TEN
Bankruptcy 57

CHAPTER ELEVEN
The Cort Parliamentary Enquiry of 1812 67

CHAPTER TWELVE
Iron production in Wales and the Black Country 72

CHAPTER THIRTEEN
Conclusion 76

Postscript 79

References 81

APPENDIX I
The wider significance of Cort's processes 88

APPENDIX II
Henry Cort's Patents 1783 and 1784 90

Index 103

LIST OF ILLUSTRATIONS

PLATES 1–8 between pages 50 and 51

1 Reversing plain rolls, 1728
2 Rolls in a slitting mill (1734)
3 Grooved rolls for rounds, 1766
4 Fontley area sketch map
5 Tombstone of Henry Cort
6 Coke refinery, Low Moor (ca. 1830)
7 Puddling furnace (ca. 1830)
8 Plate rolling and shingling mill (ca. 1830)

Frontispiece   Bas-relief of Henry Cort in Hampstead Church

Introduction

ACCORDING TO his epitaph Henry Cort was born in 1740 and died in 1800. His life spans a period of rapid technological change, not least in the iron industry, and his patents of 1783 and 1784 for the rolling and fining ('puddling') of wrought iron are to be considered important landmarks in the industrial development of Britain at the time of the Napoleonic wars. For some years I have been gathering information as a basis for a short book on Cort, a task given some greater urgency by the approaching bicentenary of his patents. To mark that bicentenary, The Historical Metallurgy Society have arranged to hold their annual Conference for 1983 at Southampton University, and I was approached with the suggestion that Dr Mott's book, which had been many years in the making, and was of a far more wide-ranging nature, could if edited provide a suitable technical history of Cort's achievements for the bicentenary. As I was familiar with the sources, and happily ignorant of the perils of editorship, I readily agreed to make the attempt, and Dr Mott's *opus magnum* duly arrived. Dr Mott was, sadly, in hospital at the time and it was felt that he would perhaps be too unwell to edit his own manuscript.

Dr Mott's manuscript consisted of two volumes, the first of which concerned itself with developments in the smelting and casting of iron. Volume II, which I received, was a history of the manufacture of wrought iron until superseded by steel. With some 24 chapters covering a very broad field, full of assiduously recorded detail, it has been necessary to select those of closest relevance to the Cort story. Dr Mott quoted Hulme: "The greatest defect of our modern histories of inventors and engineers is that they have for the most part been written by men whose scholarship has not atoned for the lack of professional knowledge of the subjects with which they have elected to deal" (Hulme, E. W., *TNS*, **1,** 192). Dr Mott is at his best in describing technical developments, and I have taken Hulme's point. The present work is about half the original length and, while retaining the most important technical description, I have attempted to prevent the book becoming technically impenetrable to the general reader. The original manuscript (volume II) was divided into 'Background' and 'Achievement' when dealing with Cort, which I have found to be a somewhat cumbersome device for a shorter work; I have

Introduction

accordingly re-arranged chapters and have opted for a chronological approach. I hope I have not done Dr Mott a disservice by eliminating much meticulous detail. He and I have lived with the Weale MS and other contemporary literature for some years and it is impossible that we should not share a regard for the achievements of Cort. The publication of this book is a timely opportunity to bring those achievements before a wider public.

Since Dr Mott finished his manuscript there have been a few additions to the literature of historical iron-making, and it is perhaps worthwhile to add some bibliographical notes.

For a general history of the iron industry the reader is referred to Ashton, T. S., *Iron and Steel in the Industrial Revolution*, Manchester, 1951; Birch, Alan, *The Economic History of the British Iron and Steel Industry 1784–1879*, London, 1967; and Gale, W. K. V., *The British Iron and Steel Industry: a Technical History*, Newton Abbot, 1967.

More specifically, Professor M. W. Flinn has shown the importance of sites rich in cordwood and water power for the development of both furnaces and forges *(Econ. Hist. Rev.,* **XI**, 1958), and this is underlined by the location of Sowley, Fontley, Bursledon and Wickham sites, described in Chapter Three. Indeed, one might argue that the southern Hampshire ironworks, hitherto ignored by the historians of the industry, should be seen as a westward extension of the Wealden iron-making region. The organisation of early forges is very well described by Alex den Ouden *(Journal of HMS,* **15**, 2, 1981 and **16**, 1, 1982) and given a longer historical perspective in Professor Tylecote's *A History of Metallurgy* (Metals Society 1976). Dr Mott's book, with its emphasis on technical change, complements an economic analysis of technological change in casting and forging practice in Charles K. Hyde's *Technological Change and the British Iron Industry 1700–1870* (Princeton, 1977) wherein is found a stimulating argument in favour of an expansion of iron output due to 'potting and stamping' rather than Cort's process of 'puddling', at least up to 1790. This argument was anticipated by Dr Mott (Chapter Twelve).

Some chapters in this book rely, by necessity, on the *Weale MS*, in two volumes, which is to be found in the British Library, Science Museum, Kensington (James Weale appears to have been secretary to Lord Sheffield). At the beginning of the work there is a printed prospectus dated 1808 of a proposed book 'An Historical Account of the Iron and Steel Trades 1777–1805', which is followed by about 200 pages of manuscript incorporating much original material relating to, and from, Cort

Introduction

and probably connected with the Petition of 1812 (see Chapter Eleven). Dr Mott has devised his own pagination, which I have observed. Where quotations at length have added to the tension and excitement of the period they have been retained (especially in Chapters Eight and Nine); where they have tended to obscure the argument they have been paraphrased and summarised (as in Chapter Eleven).

Dr Mott's book starts with an account of changing forge practice from the beginning of the eighteenth century. It might help the general reader to provide the following short background to iron-making processes, so that the new techniques can be put into their context.

Iron occurs naturally in ironstone. To remove the iron from the ironstone the latter needs to be smelted in a blast furnace. In the early eighteenth century this was usually a square stone structure about 25 feet high with an interior lining of stone. In order to smelt the iron the temperature would need to reach about 1400°C at that period, and this was achieved by use of a charcoal fuel aided by the blast of bellows. At intervals the unreduced ore and impurities floating on top of the molten iron would be drawn off as 'slag' and the molten iron poured from the base of the furnace into prepared sand-beds. The long stream of cooling metal was called a 'sow'; smaller moulds of more manageable size running from the 'sow' were called 'pigs'—hence 'pig iron'. Iron from a blast furnace could be used for castings, e.g. cannon, or stored in the form of (cast-iron) pigs. Cast iron contains 3–5% carbon, making this iron hard but brittle.

In order to make malleable, or wrought, iron further processing was required in a forge. The forgemaster would buy in pig iron, which would be fined to reduce the carbon content to less than 1% in a finery-forge, again fuelled by charcoal. In conjunction with a chafery and shingling hammers requiring water-power, bar iron would be produced for sale to the smithing trades.

During the early eighteenth century attempts were made in the casting branch of the trade to employ coke instead of charcoal in the blast furnace; this would lessen dependence on charcoal supply if an increase in output were required, or possibly improve the quality of castings. For the forgemaster this would present new problems. Coke-smelted pigs would be contaminated by sulphur from the coke, so that fined iron would break instead of bending under the shingling hammer; in addition, there would be an increase in undesirable silicon. The forgemaster, attempting to

Introduction

increase output, was restrained by charcoal fuel supplies, while the substitution of coal or coke as a fuel would compound his difficulties because of the double need to remove sulphur. The attempts to find an acceptable method for the employment of coal or coke in the fining process at the forge are the starting point of Dr Mott's book.

PETER SINGER

Fareham, Hants
September 1982

Preface

HENRY CORT'S processes for making wrought iron (with coal instead of charcoal as the fuel) and rolling to bar (instead of finishing with slow forge hammers) saved England from defeat by blockade in the Napoleonic wars—for in 1790 two-thirds of the wrought iron used in Great Britain was imported from Russia and Sweden. After the battle of Waterloo and for a period of 50 years Cort's processes enabled Great Britain to make more wrought iron than the rest of the world, whilst British operators introduced his processes (and that of the Darbys), with English railways, into most European countries.

E. W. Hulme, Librarian of the Patent Office, said in the first volume of the *Transactions* of the Newcomen Society—a society formed to overcome the defect of which he spoke—"The greatest defect of our modern histories of inventors and engineers is that they have for the most part been written by men whose scholarship has not atoned for the lack of professional knowledge of the subjects with which they have elected to deal".

The story of Cort's achievements has suffered from the defect of which Hulme wrote. There has been a general lack of appreciation that the conditions for the validity of a patent, when Cort took out his in 1783–4, were different from those of the nineteenth century and the errors of Smiles and Percy in this respect have been perpetuated by later writers; yet the validity and novelty of Cort's patents are beyond question. The story of Cort's misfortunes has, of course, been well told by Samuel Smiles but these misfortunes created incorrect estimates of production which became a further corruption of the story. Intense pity for Cort's misfortunes has embroidered the story with innuendoes of political malpractice which were unfounded. There were also, because of the free use of Cort's patents (seized for a debt owing to the Crown by Cort's backer), falsifications of the truth by those who made fortunes by using Cort's processes. Finally, Cort's processes (and Darby's coke blast furnace) were introduced into Europe as 'the English methods' so that foreign users often failed to appreciate the part played by Henry Cort in their development.

Much time has been spent in correcting the metallurgical inexactitudes, the verbal myths, the emotive misrepresentations and the falsifica-

Preface

tions, deliberate and unconscious, which have unworthily cluttered the truth. The story has been told from original sources, some of which (based on Minutes of Henry Cort himself) were sent to James Weale by Cort's sons to disprove the falsifications of the 1812 Parliamentary Enquiry. Sufficient reliable statistics have been found to give a quantitative basis for the new assessment of the success of Cort's processes.

The keepers of the archives at the Public Record Office, the British Museum, the National Maritime Museum and the Science Museum Library made the author's searches there pleasant experiences. The County Archivists at Kendal, Kingston-upon-Thames, Greater London and the City of London (Guildhall) were most helpful. The documentary assistance of the Public Librarians at Birmingham, Edinburgh, Portsmouth and Newcastle upon Tyne was of great assistance. The Rectors or churchwardens of Holy Trinity, Gosport; Holy Trinity, Kendal; Crowhurst, Surrey; St Olave's, Hart Street, London; St Giles-in-the-Fields, Camden; Standon, Hertfordshire; and Titchfield in Hampshire were generous in the time spent in searching the registers in their charge. The author has been compensated for his deficiency of local knowledge by Mr E. W. Nicholson of Messrs Camper and Nicholson, a firm which took over part of Cort's Gosport shops as a mast house; by Dr L. F. W. White, who has written the *Story of Gosport*; by the County Archivist at Winchester; and by the Archivist at Palace House, Beaulieu.

A year's work attempting to unravel the mystery of the parentage of Henry Cort ended in the conclusion that he was illegitimate; a conclusion which, despite much circumstantial evidence, could not be *proved*. The result appears as a few pages; had it been otherwise the list of acknowledgements would have to be much extended.

The goodwill the author invariably experienced in his enquiries is something he will always remember with gratitude. He cannot but think that this was, at least in part, a realisation of the worthiness of his task. The author has reason to believe that, at long last, the achievements of Henry Cort will be appreciated in their true merit and that it will be seen that his obscure birth in Lancaster was 'of momentous destiny'.

R. A. MOTT

CHAPTER ONE

The search for new finery processes

THERE WERE many unsuccessful attempts in the seventeenth century to devise alternative methods for fining[1] pig iron to wrought iron without the use of charcoal. Most of the patents of the seventeenth century included the smelting of the ore as well as the fining of the product, both processes to be carried out using the reverberatory furnace. The early eighteenth century processes are associated with the names of William Wood (1728), his son Francis (1727) and his associate Kingsmill Eyre (1736). The second half of the century saw further attempts by the Cranage brothers (1766), John Roebuck (1763), John Cockshutt (1771), Wright and Jesson (1773), Peter Onions (1783–4) and Henry Cort (1783–4). William Wood's process, including both smelting and fining, was tried out at the King's expense at Chelsea in 1731 and proved a dismal failure; the final iron, being red-short, broke under the hammer. Inventors were not discouraged and continued their efforts to fine pig iron to wrought iron in a reverberatory furnace until success was finally achieved.

Red-shortness, the inevitable consequence of heating iron ore with coal in a reducing atmosphere, was due to the release of sulphur as hydrogen sulphide, which was readily taken up by the iron to form iron sulphide. It will help the reader if it is noted here that only two processes using coal as the fuel attained any real success in fining pig iron to wrought iron. One was the process of heating the pig iron in pots so as to preserve it from the effect of the sulphur in a coal-fired reverberatory furnace, the final shaping of the wrought iron to bar being effected by the use of grooved rolls, a less drastic process than beating under a forge hammer. The pot process, which Cort's puddling and rolling process was to eventually replace, had by 1788 been applied to produce as much wrought iron as the old charcoal finery process.

1

William Wood I (1671–1730) died before the royally-sponsored test at Chelsea of his method, leaving a family of six boys and seven girls.[2,3] In his will William I, still convinced of the value of his patent for "Making of Iron with Sea or Pit Coal", left his leases, forges, furnaces, mills, etc. to two sons, William II and Charles, and to his son-in-law William Buckland. The income from his estate "chiefly consisting in Iron, Copper, Lead and Tin Works and Mines" he wished to divide into 100 parts or shares, 10 each for his widow and eldest son William II. Four of the remaining sons (excluding John) and his seven surviving daughters were each bequeathed five parts. From the remaining 25 parts his son Charles and his son-in-law William Buckland were each to retain £15,000, the remainder to be divided between his widow and William II. To John himself, from whom he may have been estranged (or to whom he may already have given his share), he left "One Shilling only".

William II remained at the Falcon ironworks, Southwark; Richard, who had an interest in Tern forge (Shropshire), died soon after his father and left his estate to Charles to pay the profits to his mother. John, though receiving nothing from his father's estate, had a forge and coal mines at Wednesbury, Staffordshire. Charles was left the works at Distington, Cumberland, where William I's chief work on his iron patent had been carried out and it was he who, by assiduous experimenting and unobtrusive development, resolved one of the problems approached in a more flamboyant manner by his father. Francis, though concerned in the first Wood iron patent of 1727, was according to family history "accustomed to pursue Pleasure rather than Business"; and the youngest son Samuel "from being one of the gayest men in England became a convinced Quaker".[3] Our attention remains with Charles and John.

Some time between 1739 and 1743 Charles Wood must have been concerned with starting, with several partners, the forge at Lowmill near Egremont, close to the rich haematite mines of Bigrigg and Cleator Moor and also to the outcrop of the Main Band, the chief seam of the Whitehaven coalfield. The first systematic entry of Charles Wood's notebooks for the Lowmill period is dated 1752, the chief MS book starting with the heading: "An account of the material transactions at Cyfarthfa in the parish of Merthyr Tydvil commencing April 11, 1766 p Chas Wood No. 1" and documenting the building of a forge at Cyfarthfa, a works which was to become the largest ironworks in Great Britain by the end of the century. Unfortunately, dates are given only periodically for the Lowmill period when certain experiments were being carried out. The

1 Search for new refinery processes

date of the most valuable entry (following experiments dated July 1753) headed "a ffurnace Erected" is therefore uncertain, but by the time Charles Wood had erected this furnace he had perfected his method of fining cast-iron scraps in fireclay pots to produce wrought iron at the rate of 8, 9 and 10 tons per week when a normal forge could not produce more than 4 tons per week. Since the first patent (curiously in the name of John Wood) for making wrought iron in pots was dated 18th May 1761, and the second patent (of John Wood and Charles Wood) was dated 26th November 1763, the date of the part beginning "a ffurnace Erected" may have been about 1760.

It is preferable to quote the claims made in the two patents before describing the experimental work recorded in the MS book. Patent No. 759 of 18th May 1761 to John Wood of Wednesbury was for "making malleable iron from sow or pig metal". It claimed that "the cast iron was refined... the fuel used is raw pit coal... these operations are therefore performed in close vessels by which means the iron is prevented from coming into immediate contact with the fuel or its flame", the closed vessels being covered pots. After heating the pots in an air furnace with a strong fire, the iron became tough and malleable and was worked into bars under the hammer.

Four examples of the procedures were given. In the first example pig or cast iron was melted in a common finery, but using raw pit coal as the fuel until it was refined in some measure and "brought near to a malleable state which the artists call bringing it into nature... the iron thus flourished I put into close vessels... and further refine it and render it perfectly malleable in an air furnace with fire of raw pit coal".

Alternatively, he melted his cast iron in a coal-fired air furnace and pounded it under water to reduce it to small grains; the granulated iron he mixed with various fluxes and cinder or scale and heated it in closed vessels in an air furnace.

A third example was to melt the cast iron into thin plates which were broken into small pieces and treated like the granulated iron of the second example.

The fourth example mentioned that if the iron was very brittle, or the coal very sulphureous, some malleable iron was first mixed with it.

In Patent No. 794 of 26th November 1763 the only differences were the use of stampers operated by a waterwheel to break the plates in the third example, and the mention of a different air furnace.

The statements of patentees of this period are often obscure but the second example of the first patent combined with the stamping procedure of the second patent is a fair statement of the practice of fining pig iron which was, by 1788, practised as widely in Great Britain as the old charcoal process at a time when Cort's 'puddling' process was about to have a still more revolutionary effect. In 1788, however, there was one alteration from the patent of 1761 for the blast-operated finery then employed in the first stage used coke as the fuel instead of raw coal.[4] This stage was later known to be desiliconisation for, in all fining processes, silicon had to be removed (as a slag) before the carbon could be oxidised.

The Wood brothers' patents for making wrought iron were based on about ten years' work by Charles Wood at Lowmill. The first entry for the Lowmill MS book is 7th August 1752. Then follow about a dozen tests dated 7, 11 and 14 August 1752, in which Cleator Moor or Bigrigg haematite or Longnor ironstone was mixed, usually, with raw coal and lime and heated in an air furnace, the product being then shingled (forged). The raw coal was usually pulverised; in one test pulverised coke was used. These tests were, in effect, a repetition of those of William Wood twenty years earlier to make wrought iron from ore and raw coal but the inevitable result is emphasised by the author's italics in an entry for Monday, 17th August 1752: "Sent to Whitehaven in order to be sent, by way of Liverpool, to Mr. Cockshutt *to take off the Red short quality*".

The next entry, dated 29th October 1752 is:

> Sent to Whitehaven, in order to be sent to Liverpool, for Mr. Cockshutt [ref. 5] Ten Hundred weight of Metal prepared from Bigrig and Longnor Iron Mines mixed & made in an Air Furn. with Pit Coal in the Proportions before mentioned. It was brought well into Nature and if Mr. Cockshutt's method of Sinking with Pit Coal makes Iron from Red Short Pig Metal I am of opinion the above Metal will produce the same. Ten Hundred might be the produce of Twenty operations. Each operation having 96 pound wt. of Iron mine which is 17 c 0 q 18 lb.

His hope that red-shortness in wrought iron (brittleness under the hammer when red hot) could be overcome was misplaced. This part of the enterprise (which was a repetition of that of his father) was impossible of achievement, though the second half of William Wood's process, the enclosure of desiliconised pig iron in pots heated in a reverberatory furnace to fine it, was perfected.

In the entry for 11th August 1752 the coal was described as "Raw Scale Gill Coal". The Main Band (the 10 ft thick seam most prized in Cumberland) outcropped at Scalegill about five miles from Lowmill with

4

1 Search for new refinery processes

the Bannock Band seam outcropping slightly to the north. These seams (like those of all coals) contained sulphur and Charles Wood was mistaken in his view that some did not, as he was also in thinking that some, but not all, slag or cinder should be drained[6] to make his iron tough and malleable. He wrote:

> The form of this finery Furnace should be, within, that is the bottom, of a Flatt surface, with a little sloop [slope], no more than will suffer the Slag or Cynder to run from the Metal as it melts . . . I do not see why the Iron should not be equally good with the Finery Iron in the Common method with Charcoal—Provided that Pit Coal be clean and ffree from Sulphur—It being well known that Sulphur is the greatest Enemy to that Metal as may be seen by mixing a small quantity with the best Tough Iron which will alter its Nature so that it may be beat in a Mortar to a powder.

From these wrong premises he went on to argue the form that a coal-fired air furnace (and his process in general) should take. The iron should be put into such a furnace "since they would be heated sooner and the slag would drain more easily from it, when the pieces in a semi-plastic form could be gathered into a loop and carried to the hammer to shingle into a half-bloom". He contended that the ordinary finery had been built with the crown raised so high above the floor that the rate of heating was slowed down but "by this New Method, the Crown may be made so low as only to allow the operation to be spread & when ready raked out by wch a greater degree of heat is obtained and the operation sooner br^t into Nature, a larger quantity of Metal got from the Oar & bro^t out much sooner".

Although he was still thinking of developing his father's 'smelting' process, all subsequent references except one refer only to the use of a coal-fired air or reverberatory furnace for fining pig or cast iron. His next entry, dated 9th May 1753, refers to the scouring of scrap to remove rust and the treating of 8 cwt 3 qr 6 lb of such iron in 13 pots, or 76 lb of scrap per pot. The entries after July 1753 refer to 35 series of tests with different clays to make large pots for this process. The pots held about 30 or (alternatively) 90 lb of scrap. The clays used were Bransay, Willimoor, Derham and others from local deposits, as well as Whitehaven rock sill. His search was for a mixture which would stand up to heating in an air furnace satisfactorily. These detailed tests are followed by a section headed "a ffurnace Erected" and, although following the series of tests referred to, seems to be of a somewhat later date, possibly 1760. He said:

> The clays for the potts are Whitehaven rock sill+Clay Got upon Willimoor Castlerigg in the following Proportions viz 5 Parts Sill, 4 parts Clay & 5 of

horse dung laid on a heap layer upon layer then cut down & mix'd well together ready for passing thro' the Iron rolls placed Horizontally three times—the first time wide enough to break the lumps small of an equal Size, the second time the rolls closer to grind it smaller, or rather squeeze and temper them together and Lastly the third time to squeeze it more to the stiff consistence of Tough Clay. After these several operations, the Clay is moulded and work'd with the hands ready for rolling with a rolling pin on a board to make the Clay of an equal thickness for the potts.

The pots were made by lining a round wooden mould of 12 inches depth, 11 inches diameter at the bottom and 12 inches diameter at the top. The mould was lined first with pieces of old sacking to prevent the clay sticking to it and then with clay; the bottom piece was put in and the joints were made. After firing, the lining of the mould was then carefully filled with a weighed quantity of cast-iron scraps and a final cover of wet clay was fitted and joined to the sides. The filled mould was then inverted onto an iron paddle slung by means of a hook and chain to a beam for the more convenient conveyance to the furnace. To those pots which were to be subjected to the greatest heat near the 'bridge'[7] extra pieces of clay were put on the top and sides and joined. Holes about ½ in. diameter were made by means of a pricker in the top and sides of all pots, the largest in the top to allow the steam and air to escape (as he said) but also (as we now know) to allow entry of the oxygen of the air to burn out the carbon.

Care was necessary in filling the pots: "when on first begin[g] this branch we were not so careful as we found afterwards was necessary—we put the scraps by hand full & shaked them down to settle the mixtures". He went on to say that they had, at first, paid a workman three-farthings a pot for filling, 12 pots being filled per day, but "now we give three halfpence a pott and get them well fill'd a greater quantity of Iron put into a pott, all clean & good so that from two, three & four ton made in one ffurnace in a Week 4 ton we tho[t] Extraordinary, we now make Eight, Nine & Ten Ton a Week". This last statement suggests a fairly lengthy experience, which is the basis of dating it to 1760; it will be seen later that John Wood may have been using the process at Wednesbury in 1754 for a production of four tons per week.

A furnace was made to hold 18 or 20 pots, each pot holding about one cwt of scraps so that one furnace charge was 17 to 19 cwt.[8] The air furnace stack (he said) should be over 40 ft high and its internal cross-section 13–15 in. square, the bottom of good firebrick, laid with thin joints of fireclay. The body of the furnace should also be of firebrick with thin joints, especially the arched roof. He advised the use of the best firebrick

1 Search for new refinery processes

"in the ffireplace & bridge which sooner burn down as it is the hottest part". The emphasis on a low crown, a bridge and a fireplace suggests the low reverberatory furnace which was to become so familiar in the nineteenth century.

In 1754 Charles Wood and his companion visited a number of ironworks, including that of his brother John, a forge at Wednesbury:

> 10th [September 1754] at Wednesbury with my brother John. He makes but 4 Ton Weekly & makes 28 cwt 2 qr into 22 cwt half bloom which is more than I can do at Lowmill. The closer the scraps are placed in the pots the less the waste will be.

The ratio of 28 cwt 2 qr to 22, i.e. 1·30, was the common ratio of pig iron to bar iron in charcoal forges. The last sentence seems to imply that John Wood's procedure at Wednesbury was the pot process similar to that which Charles was developing at Lowmill.

Marchant de la Houlière,[9] in August 1775, saw John Wood's ironworks at Wednesbury, where Wood also owned several coalworks, at two of which were 'fire-engines' of 56 and 76 in. diameter, the latter being then the biggest in the country. Scrap wrought iron, brought into the works by canal, came from northern Europe; it was filled into 20 crucibles or pots made from local clay and charged into one of two reverberatory furnaces. The pots were heated until they broke when the scrap in each was found to have agglomerated into pyramidal heaps. Each was welded to a heated iron bar, or porter, 4 ft long and "given the final shape of a loop under the hammer. Then five of six of these porters and their loops were heated in a chafery using raw coal and forged into bars under a hammer of 6 or 7 cwt". The wrought iron was said to be equal in quality to the best Swedish iron, brought the same price, and was much used for making the barrels of muskets.[9]

The *Wood MS* gives no further details until Charles Wood and others from Cumberland went to Cyfarthfa in April 1766. There they built a forge with a clay mill (with rolls to prepare the clay for making pots), a 'flourishing' house (for the pot furnaces), a stamper mill (to break the thin plates of iron from the first melting of pig iron cast onto the floor to solidify) and the forge hammers. This Cyfarthfa forge was substantially complete when the records end in May 1767 when a blast furnace 50 ft high and 36 ft square was being built. The forge and furnace were built at Llwyncelyn, north of Rhyd-y-car on the right bank of the river Taff. The forge Wood built was probably brought into operation in 1767 or early

7

1768 in what were later called the Ynysfach works; the main Cyfarthfa works were later erected more to the north towards the junction of the two branches of the river Taff.

Charles Wood was acting for Anthony Bacon & Company when he built Cyfarthfa forge. Plymouth furnace (the partners of which traded under the name of William White & Company) supplied the castings for Cyfarthfa forge; by July 1766 Anthony Bacon & Co. had acquired the three original shares of Isaac Wilkinson as well as the two in the name of John Guest for Plymouth furnace. It seems likely that the remaining shares were later acquired by Anthony Bacon who then operated Plymouth and Cyfarthfa furnaces (and Cyfarthfa forge) as one unit.

In a letter from John Cooke of Kilnhurst forge, Yorkshire, to Henry Cort dated 21st October 1788 he said:[10] "I am informed that, at Cyfarthfa, one forge is employing your method and other the Shropshire mode of Fineries with coal, and Stamping and Balling as it is there called". The 'Shropshire mode of Fineries', in fact, would be the Charles Wood pot process, still being carried out at one forge at Cyfarthfa (Ynysfach). In a letter written by Henry Cort to Richard Crawshay dated 25th October 1787[11] he recorded a visit to Penydarren forge when one of the Homfray brothers (probably Samuel) said that "whilst the others kept on stamping he should", an indication that Charles Wood's process was then being applied at Penydarren. Finally, we have the statement made by David Mushet[12] that in 1788 "there was at this time 69 melting refineries for working with coke and stamping each of which was supposed capable of using as much pig iron as would make weekly 5 tons of bar iron 60×5×52 yearly 15,600 of bar iron in the new way".

The Cranage Process (1766)

Whilst Charles Wood was building Cyfarthfa forge in 1766 he was informed by Isaac Wilkinson that the Coalbrookdale company had taken out a patent for making wrought iron from pig iron using coal as the fuel. This was Patent No. 815 of 1766[13] by Thomas and George Cranage which stated:

> The Pig, or Cast Iron, is put into a reverberatory or Air furnace built of a proper construction and without the addition of anything more than common pit coal is converted into good malleable iron, and being taken red hot from the reverberatory furnace to the forge hammer is drawn into bars of various shapes and sizes according to the will of the workman and we the said Thomas Cranage and George Cranage do hereby aver and declare that the above is a full, true and perfect description of our said invention.

1 Search for new refinery processes

Unfortunately, posterity has not found the 'full, true and perfect' description full enough and it has usually been assumed that in some way it anticipated Henry Cort's 'puddling' process. Isaac Wilkinson told Charles Wood that it was "putting it [i.e. the pig iron] into an air furnace upon a sand bottom and suffering it to remain there until brot into nature then remove it into another air furnace and further refine it". Charles Wood commented in his Journal that this was the process of one Woodhouse of 1731, that a good iron was made, but that the waste of metal was considerable and that the make per week was small. He also recorded that when he himself had tried the method the sand bottoms did not last more than two or three days and that the iron stuck to the bottom.

A further detail of the Cranage process was given in an affidavit of 16th February 1785 by Henry Foxall, Henry Cort's assistant, who stated that in a visit of his to Ketley on 29th October 1784 he had been told that in the Cranage process "the bottom of their furnace was cast metal with a stream of water running under it and when the pig iron they put in the furnace was melted they let it chill and then turned it over for the air to come to the other side". This implies that the chief oxidant was air which was usually slow in action and wasteful of iron. Foxall also related that he was informed by Thomas Cranage that he and his brother George (then dead) received £30 and "were to have had more if the process had answered but it did not . . . it failed however and was finally given up". Another statement in the same affidavit was that "the reason why they left off Cranage and adopted the granulating method was that the iron produced by the former was not fit for nails".[14]

The importance of the Cranage patent of 1766 has usually been exaggerated, an opinion probably based on a letter of Richard Reynolds to Thomas Goldney III dated 25th April 1766:

> Thomas came up to the Dale, and, with his brother, made a trial in Thos. Tilly's air-furnace with such success as I thought would justify the erection of a small air-furnace at the Forge for the more perfectly ascertaining the merit of the invention. This was accordingly done, and trial of it has been made this week, and the success has surpassed the most sanguine expectations. The iron put into the furnace was old Bushes, which thou knowest are always made of hard iron, and the iron drawn out is the toughest I ever saw. About 1¼ inch square, when broke, appears to have very little cold-short in it. I look upon it was one of the most important discoveries ever made, and take the liberty of recommending thee and earnestly requesting thou wouldst take out a patent for it immediately.

Samuel Smiles, who quoted this letter, added in a footnote:[15] "We are informed by Mr. Reynolds of Coed-du, a grandson of Richard Reynolds, that 'on further trials many difficulties arose. The bottoms of the furnaces were destroyed by the heat, and the quality of the iron varied. Still, by a letter dated May, 1767, it appears there had been sold of iron made in the new way to the value of 247*l*. 14*s*. 6*d*.'" This, however, only represents about 16 tons in one year.

Most of the processes inventors were to propose met with some success but it was probably difficulties such as a low rate of production, a high conversion ratio of pig iron to wrought iron and the problem of maintaining the furnace bottoms which finally caused all except Charles Wood's process (and a modification called 'piling') to be abandoned before Henry Cort's process dominated bar-iron making.

Roebuck's Process (1763)
Another process of bar-iron making, using coal as the fuel, was that of John Roebuck of Birmingham, Warwickshire, 'Doctor of Physick' (Patent No. 780 of 19th February 1763).[13] Until recently there was little in the records beyond the bare statements of this patent specification of Roebuck (who was, of course, the founder of Carron Ironworks) for whose ambitious plans a process for refining cast iron to wrought iron using coal would have been of great value. He claimed that pig or cast iron was melted in a hearth which was heated by bituminous coal with the blast of bellows and that he did "work the metal until it is reduced to nature, which I take out of the fire and separate into pieces: then I take the metal then reduced to nature and expose it to the action of a hollow pit coal fire heated by the blast of bellows until it is reduced to a loop which I draw out under a common forge hammer into bar iron".

This specification is no more illuminating that that of the Cranages, the expression 'reduced to nature' being notable for its obscurity of meaning. The only clear deduction which can be made is that Roebuck used a bellows-operated furnace and not a natural draft furnace. His 'hollow pit-coal fire' was another obscure expression which occurs in patent claims of the period and became common for what was termed the 'coke refinery', in which was carried out the preliminary oxidation stage of desiliconisation in the Wood process as practised in 1788. The term has been given some meaning by G. R. Morton who showed[16] that the hollow coal fire was a feature of a chafery using coal (coke) when, certainly about the middle of the eighteenth century, a coal-fired chafery

1 Search for new refinery processes

was common for the final stage (after fining, using charcoal in the finery). The hollow coal (coke) fire was formed in the morning by stacking large coal in two walls and an arch and converting it to coke by firing and operating the bellows, though the structure probably owed its stability to fused hammer scale. In effect, the forgeman created an arch of coke below which he could put his ancony for reheating before he forged it to bar, the arch or hollow coal (coke) fire giving a reducing atmosphere.

It seems possible that Roebuck was, in fact, using a *coke* fire in his finery and a hollow coal (coke) fire in his chafery, only the first stage of which might be novel at the time, but which later became the standard practice. Like many other patentees he was unduly optimistic after his first trials but Campbell has shown, from the records of the Carron Company, that the high hopes of the initial trials were not sustained. In July 1762 it was claimed: "excellent bar iron made with Pit coal without a grain of Wood coal" and John's brother Benjamin, in Sheffield, claimed to make satisfactory wool shears from it and "as it will do that, it will answer any purpose of the Sheffield Manufactory exceeding well upon which you may depend". The filing of the patent was accelerated by the fact that a Carron forgeman who had been a hammerman at Cockshutt's forge at Wortley, Sheffield, and who had seen the trials at Carron, announced his intention of returning to England. The Roebucks had, however, been over-optimistic and Campbell concluded that "subsequent events proved that he had failed to make the advance originally thought".[17]

Subsequently at Carron bar iron was obtained from the Baltic, but in 1784 a modified process with an intermediate stage using *charcoal* was being used. Gascoyne said "of the various methods of converting Pigs into Bar Iron lately tried at Carron, that of bringing it into Nature with Pit Coal, afterwards sinking it with Charcoal and drawing it in the Hollow Fire has hitherto succeeded best as to cheapness".[18]

Cockshutt's Process (1771)
On 27th August 1771 John Cockshutt of Wortley ironworks, near Sheffield, took out Patent No. 988.[13] This was of the old traditional type claiming that, starting with iron ore and without using anything but a common finery, he could produce cast iron and refine that cast iron to wrought iron in one stage. If, however, he were only concerned with fining pig iron he used a finery with pit coal (coke?) and when the pig iron was nearly melted he took it out and completed the melting in a charcoal fire; this process is the only one having any real chance of success, but since the

final fining was done in a charcoal fire it was not very notable, and was the same as that referred to as being used at Carron in 1784.

Wright & Jesson's Process (1773)
Patent No. 1054 of 2nd December 1773 was in the name of John Wright and Richard Jesson of West Bromwich[13] for a process in four stages. In the first, cast iron mixed with scale or cinder was heated in a normal bellows-operated finery but using pit coal "or coaks" instead of charcoal. In the second stage the product was taken out in lumps and beaten to plates under a large flat stamp and the plates were broken into small pieces under a round stamp. In the third stage, having removed the small particles, supposed to be sulphureous, the product was "cleansed of sulphurous matter" by washing in a rolling barrel. In the fourth and final stage, the washed residues were heated in a common air furnace "in pots or otherwise" with a fire made of pit coal "or coaks" and then shaped to bars using a common chafery.

Marchant de la Houlière visited the West Bromwich works on 14th August 1775 and saw a "refinery furnace" charged with "desulphurised" coal (coke) and 4 cwt of pig iron and subjected to intense heat until it became "heaving loops" (*loupes*). Each was transferred to an adjacent thick cast-iron plate and flattened by a forge hammer into a cake 1–1½ in. thick which, when cold, was broken under the hammer. The pieces were washed to remove coke particles and when dry were heated in clay crucibles 12 in. high, 10–11 in. diameter (like Charles Wood's) and 20 'pots' were put into a reverberatory furnace.

The furnace was lighted and after 4–5 hours the pots were all broken and the iron particles in each mingled into a shapeless block which was covered with a heated iron plate and welded into a single loop or slab with three of four blows, between which it was turned. These slabs were heated, four or five at a time, in a chafery and shingled into bars. The product was said to be hard but not very flexible and therefore difficult to work but was suitable for cart tyres, hinge iron and for making large nails. The make was 150 tons per year or 3 tons per week, a smaller production than from the Wood process.

On 24th August 1784 James Watt wrote from Birmingham to Sir John Dalrymple: "As to Mr. Cort's process . . . it is not practised in this neighbourhood, the iron having been mostly made by Wright and Jesson's process . . . by making stampt iron and potting".[19]

It is difficult to see any method in this patent different from that used

1 Search for new refinery processes

by Charles Wood, except for the alternative use of "coaks". The process as described by de la Houlière is interesting as being the first known use of the coke-fired refinery furnace for the first stage; this refinery furnace was later applied with success to the Cort process.

In a later patent (No. 1396 of 6th March 1784), after beating to cakes in stage two, the cakes or plates were piled one upon another "without pots" and heated in a coal-fired reverberatory furnace and finished as before. This 'pile of plates' method was in use at Horsehay in 1798–9.[20]

Peter Onions's Process (1783–4)

Henry Cort's basic patent (No. 1420) for using a coal-fired furnace for fining sow or pig iron to wrought iron was dated 12th June 1784.[13] On 1st September 1783 Peter Onions of Merthyr Tydfil was granted Patent No. 1370 for ostensibly the same purpose which, on 25th March 1784, was extended to Scotland. Peter Onions's patent, as well as that of the Cranage brothers of 1766 and the common British practice using the Wood brothers' patents of 1761 and 1763 or Wright & Jesson's 'pile of plates' were to be used in later controversies by the ironmasters to argue that Cort's process was not novel. It is desirable, therefore, to outline the claims made in these patents of 1783–4.

In his first patent Peter Onions was described as of Merthyr, Glam., and in his second as of Cyfarthfa, which is in the parish of Merthyr, but later records show that he was then at Dowlais ironworks where he was attempting to develop his process. The unusual name of Onions was common in Shropshire; a Thomas Onions was described as a master collier of Dawley in a Chancery case of 1754. Thomas Onions went from Coalbrookdale to Carron whilst Daniel Onions, perhaps the brother of Peter and working with him at Dowlais in 1789, could be the person of the same name said by Tredgold to have 'executed' the Iron Bridge. In fact, Peter Onions made his first trials at Ketley ironworks, Shropshire, as two letters from William Reynolds to Henry Cort show:[21]

> 17 January 1784 I am much chagrined that Peter Onions has not yet been able to succeed in the furnace, he is going to try in one built some time since by the Dale company which I hope will prove more propitious to his wishes.
>
> 17th February 1784 I am sorry to say Peter Onions has not succeeded & has with us entirely given up the point, however he seems confident of succeeding elsewhere, but this I doubt, unless he hits upon some other expedient—I am on the contrary as much pleased to hear of the great

13

success of thy experiments, and sincerely hope they will be attended with every advantage to the ingenious author that he can wish.

In his patent Onions described how he melted metal in one furnace and then transferred the molten metal in ladles to a second furnace which had a firegrate, a door for firing and a door for the removal of ashes in the ash-pit and a hearth. A blast of air from bellows or blowing cylinder was discharged below the grate and in the bottom of the ash-pit was a trough with flowing, or a cistern with stagnant, water. Since the fire-place was "filled with fuel or pit coal, coaks or wood charcoal" this would become, in effect, a generator of producer gas which would require over-fire air for its combustion. There was a pipe discharging cold air over the metal in the hearth to "excite a kind of ferment or scoriafication in the . . . metal" but it would also be essential to enable the producer gas to burn. The workmen had then to stir or turn the metal when it would "discharge or separate a portion of scoria or cinder from it and then the particles of iron will adhere and separate from the scoria, which particles the workmen must collect or gather into a mass or lump, and then shut the door and heat the mass until the same becomes a white colour" when it could be shingled and forged as in the old process.

John Percy, having given a long and correct extract from Peter Onions's patent, concluded:[22]

> The process specified by Onions was essentially that of puddling and nothing can be more graphic than the description which he gives of the 'coming to nature' of the iron, and the formation of the cinder; and I can hardly conceive that such an account could have been drawn up except by an eye-witness of the process in operation. Cort's claim to be regarded as the original inventor of puddling is thus further invalidated by the specification of Onions's patent.

It is difficult to understand why Percy wrote this passage for his extract of Onions's patent is complete so far as the operations are concerned, the only part (other than legal phrases of no present significance) which he omitted from the Blue Book of Onions's patent being the illustration and description of the forge hammers. The expression 'coming to nature' was not used by Onions though the separation of cinder was (as has been requoted above). John Percy seems to have had the same difficulty as had the contemporaries of Cort in understanding exactly what Cort's rivals did, and seems to have assumed that Onions's statement "when the metal became less fluid and thickens into a kind of paste which the workmen, by opening the door, turns and stirs with a bar . . . and then closes the

1 Search for new refinery processes

aperature again" must be equivalent to the expression 'brought into nature' (which was certainly used by Cort) or to 'puddling' (as Cort's process was described in the last decade of the eighteenth century).

It is important to appreciate how difficult it is to understand the language used by contemporary forgemen and to remember that metallurgists argued about the reactions in Henry Cort's process for a hundred years. Nevertheless, in order to understand the relevance of Cort's processes it is necessary to understand the claims of Peter Onions, the Cranage brothers and the Wood brothers whose patent claims, and processes based on them, were held to anticipate the claims of Cort.

Schubert implies in his description of the finers' process in the traditional charcoal finery 'from about 1500–1800' that the expression 'come to nature' is ancient, for after the pig iron had been melted down in the bellows blast:[23]

> The metal melted down a third time and formed pasty lumps at the bottom which were left for some time in the slag-bath. Then the finer gathered the lumps and kneaded them into a ball termed a 'bloom' or 'loop'... The whole process of melting, refining and balling took one hour. Success was judged by sounding the metallic mass with a ringer. At first the particles of slag and iron adhered tightly to it and had to be knocked off with a hammer. As soon as they began to adhere less, the finer knew that the metal had begun to change into malleable iron or, to use a phrase commonly used by the early finers, had 'come to nature'.

Schubert (although giving what was perhaps the first intelligible meaning of the phrase 'come to nature') does not give a date for it and it does not occur in the description of the Yorkshire ironmaster Pashley of 1760 which is the chief source of his description. The author has not seen the expression in contemporary descriptions of the charcoal finery but has found it in the writings of Charles Wood (1753), in the patents of John Wood (1761), of John Roebuck (1763) and of Henry Cort (1784); it seems to be an expression appropriate only to a coal-fired air furnace.

Before assessing Henry Cort's contribution to the improvements of the period, it is well to outline his background and explain how it was that his processes were developed at Fontley, near Fareham in Hampshire.

CHAPTER TWO

The background of Henry Cort: Navy Agent

IN 1789 Jane Cort of Lancaster, spinster, the youngest and last surviving daughter of Henry (Cort) of Kendal, Lancashire, left a legacy in her will of £100 to her 'cousin' Henry Cort "late of the Navy Office Crutched Ffryers London but now of Gosport in Hampshire, Gentleman" and a similar legacy and her best clothes to "his sister Jane Cort of Standing in Herefordshire [*sic*], Spinster". Standon, Hertfordshire, was to yield the first record of the existence of Henry Cort:

> 9 and 10 May 1763: Draft lease and release, Henry Trott of Standon, farrier, to Henry Cort of Crutched Friars, gent., farm in Colliers End for the sum of £825.[1]

> 10 May 1763: Surrender of Henry Trott and Amy Cowell to Henry Cort, and Cort's admission, copyhold of the above farm.[2]

The calendared records of marriage licences often serve as a guide to marriages; E. W. Hulme found in those of Surrey[3] the record of a licence for Henry Cort's first marriage in which both applicants stated their ages to be 22. The record of the marriage at Crowhurst, Surrey, is as follows:

> Henry Cort of this p^h gentleman, & Miss Elizabeth Brown of the p^h of St Giles in the Fields, in the County of Mid^x were married in this Church by Licence this (24th) day of April 1764 by me Wm Hoggart, Minister.
>
> This marriage was solemnised between us in the Presence of
> H. Charouneau (clerk)   Hen^y Cort
> James Gatland   Elizabeth Brown
>   now Cort

As the previous record of 1763 shows, Henry Cort was already in residence at Crutched Friars, London, and was of sufficient substance to purchase a farm and to style himself Gentleman. The Crowhurst registers during the period 1749 to 1812 show that the style 'gentleman' occurs only for Henry Cort. The style 'Miss' for Elizabeth Brown could be 'M^{rss}' (Mistress) and is almost unique in the records for the same period, and

16

2 Henry Cort: Navy Agent

although nothing more is known of Elizabeth Brown the unusual style suggests she may have been of some standing or wealth.

The next record of Henry Cort appears in the London directories and it is convenient to record here all the entries that occur for the period 1765 to 1775. The Directory published most regularly was that of Henry Kent. According to Goss,[4] Kent's Directory of London was published every year from 1763 to 1775 and gave, in addition to the names of the Lord Mayor and Aldermen, those of merchants and directors of companies. Copies of these directories in the Guildhall Library, London, record:

(1765)   Cort, Henry—agent, Crutched-friars
(1767 to 1769)   Cort, Henry—agent, Crutched-friars
(1770 to 1772)   Cort, Henry—agent, No. 4 Gould-square, Crutched-friars

The details from the directories may be compared with a statement (which may have been derived from Cort himself) in the Weale MS:[5]

> He was extensively engaged in business as a Navy Agent, in Gould Square, Crutched Friars, London, in the year 1768 when he married a niece of Mr. Attwick, who then held a contract with Government for supplying Portsmouth Dock Yard with mooring chains and other Iron naval stores.
>
> In 1772 Mr. Attwick assigned his contract to a Mr. Morgan. Mr. Morgan, to whom Mr. Cort advanced, at various times, considerable sums of money to enable him to carry on the concern and, in consequence of these pecuniary transactions and the embarrassed state of Mr. Morgan's affairs, Mr. Cort found it expedient in 1775 to relinquish his agency business and to settle at Gosport to take upon himself the management of the Iron Works and the supply of the Contract, which had been assigned to him by Mr. Morgan.

There is also the statement of Thomas Webster[6] for this period made when Henry Cort's son Richard was still living, the most probable source of the information:

> About the year 1765, Henry Cort had established himself as a navy agent in Surrey-street, Strand, and in 1768 married Elizabeth Heysham the daughter of a solicitor in Staffordshire, and a steward of the Duke of Portland ... After ten years of success, during which period he amassed a considerable sum of money, Henry Cort was induced, about the year 1775, to relinquish the Navy agency.

One assumes that Cort's first wife must have died. The second marriage of Henry Cort in 1768 to Elizabeth Heysham is confirmed by an entry in the Giles-Puller collection:[7]

> 16 March 1768: Copy marriage settlement: Elizabeth Haysham of London, spinster, and Henry Cort of Crutched Friars, London, gent.

Henry Cort: the Great Finer

In this marriage settlement Henry Cort settled on his future wife the farm and £2,000. Her own marriage portion is given as £3,000. Elizabeth Heysham was the daughter of Thomas Heysham, the steward of the Duke of Portland moiety of the Earl of Southampton's Titchfield estates, Hampshire, from 1731 to 1765 when the Duke sold this estate. Her uncle, William Attwick, was a prominent resident in Gosport and supplied ironmongery to Portsmouth Dockyard from 1760 to 1772. The Weale MS is supported by Portsmouth Dockyard records, the Gosport firm having the style of Messrs Attwick and Morgan in 1772, becoming Thomas Morgan in 1774. One assumes that Attwick may have died in 1774, the Gosport works then being inherited by Cort's wife.

The births of five children to Henry and Elizabeth Cort II between 1769 and 1775, when they lived at Crutched Friars, are recorded in the registers of St Olave's, Hart Street, adjoining Crutched Friars. The list of Henry Cort's children has been given by Webster[6] and Hulme.[8] Hulme had the registers of Holy Trinity Church, Gosport, examined and found six children born to Henry and Elizabeth Cort between 1778 and 1786. To these he added three others born 1788–93 to William and Elizabeth Cawte and baptised at Holy Trinity, Gosport, though there may be a confusion here with another local Gosport family.

The following list is based on Hulme but is limited to the 13 children enumerated by Webster and confirmed from other sources:

CHILDREN OF THE SECOND MARRIAGE OF HENRY CORT

Name	Date of birth	Register
1. Henry Bell	29 December 1769	
2. Coningsby	12 November 1770	
3. William Thomas	16 December 1771	St Olave's, Hart Street
4. Eliza Jane	21 April 1773	
5. Harriet Ann	22 April 1775	
6. John Harman	— January 1777	Unknown
7. Maria	13 June 1778	
8. Charlotte	12 October 1779	
9. Frederick John	10 October 1781	Holy Trinity, Gosport
10. Caroline	18 January 1783	
11. Richard	20 April 1784	
12. Louisa	21 June 1786	
13. Catherine Frampton	21 February 1790	Unknown

The sixth child was born after the family had left Gould Square, Crutched Friars, for Gosport, but perhaps had not settled in their final house. The registration of his baptism might have been in the Fareham registers but these have not survived for

18

2 Henry Cort: Navy Agent

the period concerned. Since in 1794, when a number of influential citizens made an appeal to the Prime Minister on behalf of Henry Cort, they mentioned 'twelve children', this reference is taken to mean that this child died in childhood.

Before moving to Gosport, Henry Cort was a Navy Agent. According to H. W. Dickinson,[9] a Navy Agent is "a Banker, and an attorney who acts for H.M. Ships as to pay, allowances, distribution of prize money, salvage, etc." The most plausible explanation of Henry Cort's activities as a Navy Agent from 1765 to 1775 is that he was primarily agent for his wife's uncle William Attwick, that she inherited the works in 1774, and that in 1775, with the extra possibilities to be expected from the outbreak of war in that year (and the complications introduced by the blowing-out of the Hampshire charcoal blast furnaces), he took over the management of the Fareham ironworks. He may have acted in respect of pay and allowances—an activity which certainly demanded the standing of a gentleman, as he was styled in 1763 and 1764 before he became a Navy Agent, and would have been familiar with the fluctuating demands of the Royal Dockyards, the chief of which was Portsmouth, and to which William Attwick was the major supplier of ironmongery.

It is instructive to analyse the demand for iron from the Navy. Iron accounted for the largest costs in a man-of-war. Swedish iron was supplied from 1761 to 1776 by Andrew Lindgren, merchant of London (who also supplied Stockholm tar for impregnating the timbers). On 18th April 1769 he contracted to supply 702 tons of iron, 275 tons of "the first sort of Orgrunds" at £19 2s 6d per ton, 137 tons of Stockholm at £17 2s 6d per ton (with a surcharge of 10s 0d per ton to Portsmouth and 15s 0d to Plymouth). The total contract was worth £14,000; 397 tons were to be supplied before Christmas 1770 so that the whole extended over perhaps three years. This would be the material used by the dockyard smiths to make anchors, chains and other large naval ironmongery.

The guns in 65- and 74-gun ships would weigh 120–150 tons and cost over £2,000 whilst the shot might account for half this sum. Mooring chains were worth 50 per cent more than the iron from which they were made and anchors three times as much.

Nails, spikes and boltstaves, however, accounted for the biggest single section in the stores required for shipbuilding and were of great variety. Boltstaves were of 1¾ to ½ in. diameter in steps of ⅛ in.; spikes (boat and ribband to fasten the rib bands) were of 18 to 10 in. in length in steps of 1 in. (with one size of 22 in). Double-deck and sheathing nails were of 9 to 5 in. in length in steps of 1 in.; sharp (rove and clench) nails were described by their cost in pence per 1000 (60d, 40d, 30d, 24d, 20d, 10d, 8d, 6d, 5d, 4d, 3d, and 2d), 20d and 10d being the commonest in use. There were also single (deck and port) nails; plate as well as wherry, round head and welding nails of 4 to 2 in. length; floor brads or battens and scupper nails. In a twelve-month period (1768–9) William Attwick of Gosport supplied 191 tons of boltstaves, spikes and nails; in 1775–6 his successor, Thomas Morgan, supplied 166 tons and in the succeeding year 208 tons. Attwick also supplied, in 1768–9, 37½ tons of mooring chains, 150 fathoms with 3 ft links at £31 per ton.

Attwick also delivered an extraordinary variety of ironmongery, some items in large numbers. In 1768–9 he supplied over 4000 locks, nearly 6000 pairs of hinges, over 2500 staples, approximately 1250 auger bits of 2½ to ½ in. diameter and the same number of scrapers, 2400 "Dutch rings" (mast hoops?), 1000 pump hooks, 500 tackle hooks, 400 boathooks, 6600 wood screws, 900 cold chisels, 624 axes and hatchets, 242 hammers, 150 marline spikes, 78 pitch ladles, 72 files (½ in. round and 3 in. square), 30 saws, 36 flesh hooks, 96 each of fire shovels, tongs and forks, 12 kettle easels—as well as some pump fittings, perhaps worth a total of £600.

It is difficult to give an accurate picture of the costs of different stores for the period they lasted is uncertain; but at least it is possible to emphasise the significance of iron goods for shipbuilding. In the following estimates it is assumed that the hemp was three years' supply, that half of the previous allocation of Swedish iron to Portsmouth was for 1769, and that the wood was one year's supply, as were the nails, spikes and boltstaves.

The estimate of nearly £8,000 of ironmongery supplied by William Attwick in 1768–9 was undoubtedly the most valuable of all contracts from Portsmouth Dockyard.[10]

The iron supplied, 288½ tons, exceeded the nominal capacity of Titchfield (Fareham) forge in 1750 (200 tons); its capacity might have been increased, or iron could be taken from store, or Swedish iron purchased. In addition, there may have been a demand from Plymouth Dockyard, which Gosport was geographically favoured to supply. It is not difficult to surmise that Henry Cort's marriage connection with the iron trade, and his knowledge of existing and potential orders for naval stores, led him to give up his lucrative calling next to the Navy Office at Crutched Friars and to move to the vicinity of Portsmouth.

20

2 Henry Cort: Navy Agent

APPROXIMATE COSTS OF DIFFERENT STORES, PORTSMOUTH, 1768–9

Wood
500 loads New Forest oak @ 38s, 50 beeches	£1,188	
487½ tons 'Petersburgh' @ 27s ⎫	943	
172½ tons Riga pine @ 33s ⎭		
70 loads of Dantzig oak planks @ £6 15s 0d	452	
21 masts of N. American white pine	1,600	
		4,183

Hemp for sails and cordage   3,207

Iron
137 tons nails @ £35	4,796	
54 tons boltstaves @ £30	1,620	
37½ tons mooring chains @ £31; 1 ton shackles	1,194	
Ironmongery	600	
100 tons Swedish iron @ £20	2,000	
Guns	2,000	
Shot	1,000	
		£13,210

CHAPTER THREE

The early history of Fareham Ironworks: Fontley and Gosport

THE REPUTATION for naval ironworks which the Fareham ironworks (Fontley mill or forge and Gosport smiths' shops) gained under Henry Cort's management was a development of a trade which had certainly flourished for two hundred years. The Portsmouth dockyards were extended by Henry VIII but their chief development was in the second half of the seventeenth century. By 1709, when the Board of Admiralty took over the powers previously vested in the Lord High Admiral of England, the area of the dockyards had been increased, in the previous 50 years, from 10 to 58 acres.[1]

To Portsmouth Harbour there was a mile-long approach with a narrow entrance fortified on both sides and near the upper end on Burrow Island on the western side was a fort, sometimes known as the Castle. Below this, also on the western side, was Gosport Green, facing which were the smith's shops and warehouse which Henry Cort later held. A ferry joined Portsmouth with Gosport. Fontley Iron Mill, 1½ miles north of Titchfield, had a connection with Gosport via the tidal creek at Fareham, in the north-west corner of Portsmouth Harbour, facilitating the despatch of its products to the associated smiths' shops at Gosport, when the fabricated products passed to the dockyards.

At the Dissolution, the lands of the Abbey of Titchfield were granted to Thomas Wriothesley who, in 1547, was created Earl of Southampton. The third earl of the second creation (1574–1624) was one of Shakespeare's patrons, and actively encouraged iron-making at Titchfield and Sowley after his release from the Tower in 1603. The fourth earl, Thomas Wriothesley, died in 1667 and though the title became extinct most of the lordship of Titchfield passed to his daughter Elizabeth who, in

3 Early history of Fareham Ironworks

1661, had married Edward Noel, Baron Noel of Titchfield (1661) and later first Earl of Gainsborough. The second Earl of Gainsborough died in 1690 leaving two daughters as co-heiresses: Elizabeth who, in 1704, married Henry Bentinck, created Marquis of Titchfield and Duke of Portland,[2] and Rachel who, in 1706, married Henry, second Duke of Beaufort.

Titchfield Iron Mill was, however, disposed of before the lordship was divided between Elizabeth and Rachel, the grand-daughters of the fourth and last Earl of Southampton. Elizabeth, the heiress of the fourth earl, then Baroness Noel of Titchfield, in 1671 made a settlement to pay her debts and to provide portions for her daughters. She leased, for "two thousand years . . . all that Iron works called Titchfield Hammer together with a Tenement, orchard and garden with a meadow and three parcels of land in Titchfield aforesaid now or late in the occupation of Roberte Greengoe or his assignes". The lease was to Anthony, Lord Ashley (who in 1672 became the first Earl of Shaftesbury), William Russell (son of the fifth Duke of Bedford who was married to Rachel, Elizabeth's sister) and Thomas Corderoy of Titchfield.[3]

This lease provides the first clear indication that the "Iron works called Titchfield Hammer" was a forge for making wrought iron. It shows also that it was leased to a Gringo(e), a family which held it for a further hundred years. Before pursuing this we should give some indications from the rent books which, though the information is indirect, serve to show that the forge had long been in operation.

The earliest mention in the records which have survived is for the lease, for 40s 0d, of a meadow "by the Iremill ponde" in 1606–7.[4] This may be the same meadow which was held in 1624–5 by Jane White, widow: "a parcel of meadow ground in Segensworth called Myllmead" for 40s 0d.[5] In 1632 widow Loffe held "a mead of the Iremill" which contained two acres and was valued at 40s 0d, whilst Henry Serred held "the mead under the Iremill Coppice" and the "Iremill Coppice 3a 1r 19p" with 32 trees was valued at 13s 0d. In the same year the "Shop" was held by Robert Gerratt for 20s 0d[6] and in 1640 a shop formerly held by a Robert Garrett was described as a smith's shop.[7]

In 1671 the "Iron worke called Titchfield Hammer" is unmistakably a waterwheel-operated forge for making wrought iron and the term "Iremill ponde" in 1606–7 can be accepted as meaning the same, the shop or smith's forge of 1632 and 1640 being a small accessory. Although we have no record of Titchfield Hammer or Iron Mill before 1606–7, it would

Henry Cort: the Great Finer

not be surprising if it was operating in the period 1550–75, the period of great expansion in Sussex.

Although the indications of the existence of Titchfield Iron Mill for the century or more before 1671 are only fragmentary, the records for the succeeding century show the continuous occupation of the Iron Mill by one family.

The occupation of the Iron Mill by the Gringoes can conveniently be shown in tabular form:

1671	'Titchfield Hammer' in the occupation of Roberte Greengoe[3]
1679–80	Roger Gringoe paid half-yearly rent of £25
1696	Roger Gringoe still in possession when the particular series of records ends
1712	John Gringoe, cousin of Roger, succeeded him from Lady Day
1740	John Gringoe paid rent of £60 for "the Hammer and Forge and Shop in the Forge"
1754	John Gringoe still held the forge[8]
1779	John Gringoe's will showed him to hold Fareham Tide Mill, Bursledon furnace(s?) and ponds and a quay at Hardley, near Fawley

John Gringoe died in 1773, while William Attwick gave up his contract for supplying naval ironmongery to Portsmouth in 1772; it was during this period that Henry Cort moved to Gosport and supervised the Titchfield (Fontley) forge.

The records of the capacity of British forges in the eighteenth century include the following entries for Hampshire (tons p.a.):[9]

	1717	1736	1750
Titchfield	140	100	200
Sowley	100	50	50

In 1717 the average British forge had a capacity of 122 tons p.a., so that Sowley was then below, and Titchfield above, the average. For comparison, there were in 1717 nine forges in Sussex (none of which exceeded a capacity of 50 tons p.a.), three in Surrey and one in Kent.[9] The two Hampshire forges were therefore the most flourishing in or near the Weald and by 1750, although Sowley had declined, Titchfield had a capacity almost equal to the British average for that year (212 tons p.a.).

Clearly there was a factor favourable for the Hampshire forges to

3 Early history of Fareham Ironworks

survive the competition that caused the Wealden forges to decline. It was not the cost of charcoal, for Heathfield in Sussex had lower costs for this in 1741 than its competitors elsewhere. The reason will become evident if we study the records of Sowley furnace and forge.

Sowley forge was in operation from 1605 to 1821, a longer period than that for most Wealden forges. Sowley furnace (which, with the forge, was 'newly erected' in 1605)[10] was blown out in 1772[11] when only Heathfield, Fernhurst, Robertsbridge and Ashburnham, of the Sussex furnaces, were still in operation[11]—so that Sowley furnace also had a life which was unusually long. The site was due south of Southampton in the manor of Beaulieu and virtually on the coast and favourable for water transport of the raw materials and finished products; but it had another advantage, for the ironstone outcropped in the cliffs to the west and was originally picked up from the beach.[11] The placenames Perrywood Ironshill and Irons Hill Walk, near Lyndhurst, west and north-west of Beaulieu, suggest that ironstone was also mined. Between Beaulieu and Sowley there were numerous copses which were, doubtless, the source of wood for charcoal making. Thus the site was particularly favourable and it is not therefore surprising that ironmaking persisted so long.

The second Duke of Montagu, whose father had inherited Beaulieu through his first wife Elizabeth Wriothesley, daughter of the fourth Earl of Southampton, leased Sowley ironworks to John White and subsequently, in 1730, to Myles Troughton, whose grandson operated the ironworks from 1756 to 1772 when the furnace went out of blast. The forge was continued by James Stairs until 1784, by Joshua Cope until 1790, and by Charles Pocock and his son Henry until 1821 when the forge was abandoned.[11] The first Myles Troughton of "Bewley in Hampshire" was co-lessee with Thomas Hall of Cranage, Cheshire, of the Duke of Montagu's Lindal and Dalton haematite mines, Furness, Lancashire, in 1729 for 16 years,[12] and Thomas Hall was transferring Sowley pig iron to the Stour forges in 1714–15 and 1737–8,[13] whilst William Ford of Newlands furnace, Furness, did the same between 1767 and 1772. It can be concluded from this that Furness haematite would be delivered to Sowley furnace, and would be another reason for its economic survival.

The fact that John Gringoe held land at Hardley by a quay on the western side of Southampton Water could be an indication that this was for the collection of charcoal from the same sources as for Sowley furnace, for Hardley and Sowley are approximately equidistant from Beaulieu and both ironworks were probably worked together by the Earls of Southampton. Subsequently the Gringoes are found in control of Sowley, furnace and forge, Bursledon furnace, Fontley forge and Wickham forge, some few miles upstream from Fontley, providing a good example of early vertical integration in the industry.[14] Since we know that the Fareham ironworks contract with Portsmouth Dockyard changed hands in 1772, and Sowley furnace was blown out in that year, it probably also marked the end of the Bursledon furnace. This realignment of the economics of Fontley forge perhaps helps to explain the difficulties which obliged Henry Cort to participate more closely in its affairs.

Although Henry Cort did not relinquish his activities as a Navy Agent at Crutched Friars until 1775, he had had an indirect connection with Fareham Ironworks for some time before, as we have seen, probably as an agent for William Attwick. The chief source of information on Cort's early association with Fareham ironworks is the Weale MS:[15]

> [Henry Cort] in the year 1768 when he married a niece of Mr. Attwick, who then held a contract with Government for supplying Portsmouth Dock Yard with mooring chains and other Iron naval stores. In 1772 Mr. Attwick assigned his contract to a Mr. Morgan. Mr. Morgan, to whom Mr. Cort advanced, at various times, considerable sums of money, to enable him to carry on the concern and, in consequence of these pecuniary transactions and the embarrassed state of Mr. Morgan's affairs, Mr. Cort found it expedient, in 1775, to relinquish his agency business, and to settle at Gosport to take upon himself the management of the Iron Works and the supply of the Contract, which has been assigned to him by Mr. Morgan.

The release of the contract by William Attwick in 1772 can probably be connected with the ending of John Gringoe's lease of Fontley forge but since the smith's shops at Gosport were primarily nailor's shops, the slitting mill at Fontley would be essential to supply nailor's rods for the Gosport smiths. On the other hand, Bursledon furnace was no longer in operation to supply charcoal pig for the finery at Fontley and Sowley pig iron was also not available. Thus the practice at Fontley required a new orientation. In effect, the alternatives were to import Swedish bar iron or scrap wrought iron for use at Gosport or to use coke pig iron (say, Welsh or Scottish) or scrap cast iron in the Fontley finery. In the event, Henry Cort used Swedish bar iron to meet immediate demands, but the use of both scrap wrought iron and scrap cast iron was to prove the basis of his greatest contributions to iron metallurgy.

CHAPTER FOUR

The development of puddling and rolling at Fareham

WE HAVE seen in the previous chapter that William Attwick, uncle of Henry Cort's second wife, ceased his connection with Fareham ironworks (Fontley forge and Gosport smith's shops) about the same time as the death of John Gringoe, whose family had operated the forge and Bursledon furnace for a century. We have assumed that Henry Cort took over the Fareham ironworks from his wife's uncle in 1772 and put in a Mr. Morgan as manager until the end of 1774 when, with Morgan getting into difficulties, Cort gave up his naval agency and himself operated the ironworks. There is then a period of five years when our knowledge of his activities is meagre; but towards the end of 1779 there is a letter of his to Matthew Boulton, of Boulton and Watt, seeking information for improving the forge and its waterwheel-operated hammer, which gives the first details of the forge.[1]

– Boulton Esq
in the care of Mr. Tho. Wilson
ChaseWater near Truro, Cornwall

Gosport
27 November 1779

Sir,
I have rec^d a Letter from Mr. Henderson, wherein he asks several Questions relative to my Mills, & to wch he requested me to give Answers to You, & that you would be Oblidgeing as to give me Your Advice.
My Mill is Undershott
The Diameter of the Wheel 12 feet ½
Fall of Water 6 feet
Width of Sluice on Penstock 3 feet & when drawn to its
    full hight is 14 Inches
The Pond when full is 8 feet (deep)
Wheel makes 25 Revolutions p Minute

Weight of Hammer 5 Cwt lifts 20 Inches & makes
100 Strokes p Minute
It is impossible to make it Overshott, as it would Drown the
Mill above, & if the Banks were raised it would not I suppose
Drive the Hammer to do any Business, for there woud be only
4 feet head, & now there is 8 feet head & the Bellows
woud go longer than the Hammer, *so says my Millwright*—but
that to your better Judgement.

  If you can make any Improvement I should be glad to
acknowledge the Obligation You may conferr on
Sir, Your most obed^t Servant
Henry Cort

Our chief source on this period states:[2]

Under Mr. Cort's management this concern was much improved; but, though the Profits were considerable, they had not been sufficient to liquidate the outstanding demands on it, in 1780, when a large sum still remained due to Mr. Cort. In that year Comm[r] Kirk of the Victualling Office appplied to him to make some new works, which the Comm[r] of that Department had occasion for at the port of Portsmouth, and, in the course of conversation, inquired whether he could make Iron Hoops; to which Mr. C. replied that he had no mill, but that he had a water situation, at Fontley, with forge hammers only, where a mill for ye manufacture of hoops might be erected. The Comm[r] strongly expressed his wishes that Mr. Cort should there establish a manufactory of Iron Hoops for Government, alledging that the Victualling Board, in making their public contracts for that article, could procure supplies only from the different proprietors of mills in the vicinity of London, who combined together at the time of tendering for the contracts, and, of course, obtained the prices which they chose to demand, a certain proportion of the whole quantity required being allotted to each proprietor, according to their private agreement; and that, the Board having no other recourse could not remedy the above. Mr. Cort stated that, he had no objection to erect a mill for the purpose, provided he were assured of proper encouragement; and, after some official correspondence, Mr. Cort entered into a Contract, to commence 1 Oct[r] 1780, to supply new Hoops for the service of the whole of the Navy, during the war, to be delivered at the Port of Portsmouth at £21 per Ton, subject to the Discount of 5 per cent, leaving ye net price at £19 19. 0d.; and, he contracted, besides to deliver one ton of new for two ton of old iron hoops, equivalent to £10 per ton for the old iron, which had frequently been sold for less than £8 and seldom more than £9 per ton. The current price of Iron, of the quality adapted for hoops, was at that time £15 per ton. Mr. Cort, acordingly, erected a Mill, at a very considerable expense, with every prospect of an adequate remuneration; but, the former contractors, exasperated at the success of the Victualling Board at breaking their combination, practised every art to enhance the price of the material, and did, in fact, soon contrive to raise it several pounds

4 Puddling and rolling at Fareham

per ton, besides throwing various obstacles in Mr. Cort's progress, with the view to ruin him, the ultimate result of which proceedings was that he sustained loss of nearly £10,000 by the Contract.

An analysis of this statement shows that the price of Swedish bar iron of £15 per ton applied to the year 1780.[3] The wars with France and Spain associated with that of American Independence, and the consequent increased demand for iron in wartime, forced the price up to £19 6s 2d per ton in 1782. Moreover, the significance of the price of Swedish iron must mean that Henry Cort made his new mast hoops from such iron and his contract price for deliveries 'during the war' of mast hoops at £19 19s 0d per ton was based on the 1780 price, so that the rise of the bar iron price to nearly that of the fabricated product would give him no profit in 1782. The reference to old mast hoops at £10 per ton meant that, instead of cash payments on delivery of the new mast hoops, the Navy Board could pay in old mast hoops which left him with a stock of these, no cash, and a loss on his dealing with Swedish iron. Assuming that he made 200 tons in each of the two years to 1 October 1782, his total expenditure on Swedish bar iron and working costs would only be of the order of £4,000. The 'nearly £10,000' mentioned as being a loss on the contract must therefore have been chiefly due to his costs in adapting Fontley Mill for his purpose.

To make mast hoops by the old process would involve rolling to plate, slitting to rod and forging on an anvil with a groove appropriate to the hoop section, shaping to a circle and welding the joint—a long and laborious process for the thousands of hoops required. It seems likely that during this period he had already adapted his mill to rolling to mast-hoop section and also welding so that he could cut out the laborious forging operations—and that it was this step that attracted the attention of Adam Jellicoe. The account continues:

> Soon after the commencement of the works under this Contract, Mr. Cort, requiring an increase of capital, entered into a negotiation for taking Mr. S. Jellicoe, the son of Mr. Adam Jellicoe, into partnership. The latter had been, for many years, whilst Pay Clerk at Portsmouth, the Agent of Mr. Cort in receiving money there due to Officers for whom he was concerned, and there had been other pecuniary transactions between them. In 1781 the terms of the partnership were arranged, and, at that time, Mr. Cort stood indebted to Mr. A. Jellicoe several thousand pounds, which had been advanced to assist in the prosecution of his undertakings.
>
> In the course of the years 1781, 2 and 3, extraordinarily large quantities of old Iron hoops very greatly exceeding the proportion of new hoops supplied [were received] so that they had not an opportunity of working

them up, in any ordinary way, without incurring a great loss. Under these circumstances, Mr. Cort entered upon a variety of experiments, with a view to the profitable conversion of their large stock of old hoops, and, at length, discovered & perfected the process, for which he obtained his first patent; and, continuing to prosecute his experiments with unremitted assiduity, he made the yet more important discoveries for which his other patents were subsequently granted, and succeeded in bringing to perfection the whole of his inventions.[4]

The stock of old iron hoops proved to be a blessing in disguise and gave the urge to remake them into hoops of good quality. The secret was in the rolling process, using the grooved rolls he had devised for making mast-hoop section from Swedish bar. As his first patent (No. 1351 of 1783) shows, he broke and straightened the old hoops, faggoted them, reheated, forged and rolled to the required section; this process removed the rust as a slag and the gentler process of rolling instead of hammering gave a grain structure which conferred a toughness to the product. Once the quality of the product had been established, a comparatively simple process like this, which doubled the value of the raw material, did not need an expert to appreciate its significance.

CHAPTER FIVE

The rolling of metals

HENRY CORT did not, as is sometimes said, invent grooved rolls but he was the first to practise and to patent the finishing of iron bars by the use of grooved rolls instead of by the forge hammer. The merit of this was that, even before Watt's rotary steam engine was applied to operate Cort's rolls, these enabled an eight-fold increase to be achieved in the output of the average charcoal forge.

The use of smooth rolls for rolling metals (including iron) to plate and even to sheet was well established by the middle of the eighteenth century; the plate could be slit to cart-tyre iron, hoop iron and nail rod. Two Englishmen, Payne (1728) and Purnell (1766), patented the use of grooved rolls for producing round iron but, since the former proposed to operate his rolls with a windmill (which would be impracticable) and the latter left little evidence of 'putting into operation', such prior disclosure would not invalidate Cort's later patent as it can be established that he, and he alone, introduced the use of grooved rolls in place of the hammer for finishing iron to bars.

To appreciate the novelty of Cort's invention in terms of contemporary practice, it would be useful to discuss the different stages of the early development of rolling processes in their proper sequence.

Gold, silver, copper, tin, lead and brass are all soft metals and can be rolled in the cold condition. Iron is much harder than all these metals and can be rolled only when red-hot.

The wire 'draw-mill' was the precursor of grooved rolls. For producing gold and silver wire hand operation was adequate, the metal plate and sheet produced by hammering being slit to small bars and then pulled through a die—a draw-plate with the aperture of the required profile—by means of a hand-operated bobbin or roll. A somewhat similar process was used for making brass and iron wire of circular cross-section, the operator being called a 'girdler' because he wore a girdle round his waist to pull the wire through the draw-plate. For all these operations the metal was cold, for even iron, if adequately fined, could be cold-drawn to wire.

The first mechanisation of this process was, not surprisingly, for iron wire, a waterwheel-operated bobbin being used to replace the girdler. A waterwheel-operated wire draw-mill (for iron and brass) was introduced into England, at Tintern, as part of Sir William Cecil's proposals for the development of English manufactures in 1568.

The Elizabethan period also saw the introduction in the Thames valley of a waterwheel-operated slitting mill for making nail rod using giant shears on a red-hot plate which had previously been hot-rolled, this being the first known English use of long rolls. Whatever the origin of Bulmer's patent for making nail rod (1578) it was, in effect, superseded by the rolling and slitting mill of Richard Foley in 1628 at the Hyde, on the River Stour. This had rotary disks tipped with steel to slit the red-hot plate (previously hot-rolled), the disks above and below the plate being set at intervals along two shafts, each separately operated by a waterwheel. This process, rapidly imitated elsewhere, made the use of such rolls familiar in different centres of nail making.

The beginning of the reign of Elizabeth I also saw the introduction into England of the rolling mill, not only for making silver strip for coinage but also for making the actual coins. Rolling mills for the copper strip required for the copper coinage were introduced at Mitcham ca. 1700, while Thomas Hale had taken out a patent for the manufacture of lead sheets "made thinne by a certain engine or rollers" in 1670 (No. 158). The process of rolling iron plate to sheet for tinning was developed and perfected at Pontypool, Monmouthshire, ca. 1695 by John Hanbury. As a contemporary account stated:[1]

> One Major *Hanbury* of this Ponty Pool, shew'd us an excellent Invention of his own for driving hot Iron (by the help of a Rolling Engin mov'd by Water) into as thin Plates as Tin

Bars of iron 24 in. long and 4 in. square were heated and passed broadside between rollers to make plates 24 in. wide and 48 in. long (a reduction in thickness of 12/1). This was the first stage and produced plate which sold in London for making pans, kettles, etc. At Pontypool the plate was subsequently reduced, in three further stages with inter-stage reheating, to sheet for common tinplate. Two further stages of reheating and rerolling were used to produce the thinnest sheet;[2] in all stages after the first the reduction in thickness was approximately 2/1. To the 'black plate'

5 Rolling of metals

of the required thickness descaling, using ammonium chloride and beating, was practised and a final second scale was first produced by heating and then removed by acid pickling to give bright sheet. This was given a final cold-rolling to produce a smoothness which could not be attained by hammering (for such thin sections cooled too rapidly) so that this development greatly improved the quality of tinned plates. This process was started at Wortley in 1743 with Welsh operators.[2]

By the early eighteenth century the practice of designing and operating plain rolls was highly advanced in England. Swedenborg said that rolling mills with slitting cutters were to be found in England, Germany and Liège, his illustration of such machinery showing two waterwheels on one side of the building (instead of on two sides as in 1683) and although this required gear wheels at right-angles to the driving shafts it enabled the rolls to be operated at a higher speed than was possible by direct drive and therefore gave a greater production. Such geared mills were called double rolling and cutting mills in comparison with single mills with direct drive. The cutting rolls consisted of cutting disks with distance pieces between. The disks were slightly concave and were either tipped with steel or case-hardened to give hardenable edges. To give the smooth surface for plate and sheet rolls, the surface of cast iron rolls was covered with steel sheet, welded on an incline. The use of harder white iron was preferred to grey cast iron, such rolls being cast round a wrought-iron shaft which had been heated to incandescence.

The plain-rolling mill of the period 1628–1728, while largely due to the advances made by Richard Foley and Thomas Hanbury, nevertheless probably owed some of its features to suggestions of unknown operators. A limit was imposed by the low power which could be developed by waterwheels of average size or which could be transmitted by wooden gears and shafts. Up to the middle of the eighteenth century there was, therefore, a limit imposed to the performance of rolling mills and they were regarded as an adjunct to, and not a replacement of, forges. The defects of such wooden fittings became more obvious in the gun-boring mills of the second half of the century. Cast iron shafts were substituted for the wooden shafts of waterwheels for such mills at Carron in 1769[3] and cast iron shafts and gearing were common when Henry Cort's patents were obtained.

Some novel ideas on rolling were either developed or at least conceived by the Swedish engineer Christopher Polhem (1661–1751) in the first half of the eighteenth century. It was at a metal works at

Stjernsund, in 1705, that he constructed the rolling mill for which he is most famed. It consisted of two slender rolls of wrought iron, covered with steel sheet, backed by two heavier cast iron rolls to prevent the slender rolls from yielding, a method resuscitated in the American Lauth rolling mill of the nineteenth century (but with three rolls) to give fast production of sheet. In 1747 Polhem wrote, at the age of 86, his *Patriotic Testament*, which was not published until 1761. In chapter 14 he said:

> Much time and labour can be saved by good rolling mills, because a rolling mill can produce 10 to 20 and still more bars at the same time as is wanted to tilt only one bar with the hammer. Thus very thin bar iron can be made which is useful for hoops and mountings of several kinds. Steel can also be rolled out for knife blades, etc. which can be easily finished by the blacksmith. The rolls can be so made that the knife-steel becomes broad and thin on both sides, or gets the same shape as blades of common swords, and these can be cut lengthwise in two parts, this giving suitable material for knives etc. *Rolls also can be made for producing quadrangular, round or half-round bars, not only for iron* but also for steel, as for all kinds of files which easily can be finished by the blacksmith.

The most interesting part of this, from our present point of view, is that italicised (by the author) and concerned with *grooved* rolls for iron, though he only said that they *can be made*. The profile rolling of hard steel (which he also said *can be made*) never replaced forging, which consolidated the steel better. It does not appear that Polhem succeeded much (if at all) with his grooved rolls (though Johannsen says he used them for key rod) for he complained bitterly about the opposition of the old master smiths. He also said that, because of economic difficulties, he was obliged to leave unused his slender rolls and other expensive machines and concluded: "Yet I willingly grant to others, who will perhaps live during more happy times, what I have not got an opportunity to use for myself".[4,5]

We are on the same uncertain ground when we look at the early patent literature of profile rolling, claims for which were limited to England, where the design and use of plain rolls had been more advanced than elsewhere. The earliest claims for profile rolling with grooved rolls had been confined to lead, the easiest of the metals to roll (1728). English patent No. 207 was granted in 1679 to Thomas Harvey, a dealer in iron, for "drawing Spanish or Swedish iron into rounds for bolts for shipping", though the 'drawing' seems to have been a finishing rather than a forming process. In 1728 John Payne was granted patent No. 505 to pass hammered iron bars "between two large mettall rowlers (which have proper notches or furrows on their surfass)" but he proposed to operate

5 Rolling of metals

his rolls with a windmill—an impracticable proposal. John Purnell (No. 854 of 1768), however, gave a picture of grooved rolls to produce rounds, the rolls having a coupling box and nut pinions to drive the second roll, which would be practicable. It will be observed that Purnell's rolls were compounded, with inserted sections for the grooves on one roll projecting into a deeper groove on the other. They were only designed for ship's bolts, rod and wire—and could not be used for rolling blooms and slabs to bar. Purnell, an ironmaster of Froombridge, Gloucestershire, was commanded to lodge a specification in the Court of Chancery so that, unusually, we have an exact specification.

All these statements, beginning with that of Polhem, do not amount to proof of the use of grooved rolls in practice. Hall[5] considered that, before Cort's practice, grooved rolls were used as a finishing stage for shapes already formed by the conventional methods of the forge.

The significant factor in a patent of invention in the last quarter of the eighteenth century was not earlier *proposals* or *principles* but contemporary *practice* and there is not the slightest doubt that Cort's *practice* was novel. Rolling in grooved rolls was mentioned in both his patents of 1783 and 1784. In the earlier one, for making new mast hoops from scrap hoops, he bundled the old mast hoops into a faggot, brought this to a welding heat in a coal-fired reverberatory furnace and forged it under a heavy tilt hammer of 8 or 9 cwt. He might pass the heated faggot through the plain rolls or a common rolling and slitting mill to express the slag and convert the iron into a fibrous and tough state but, finally, "bars, bolts, half-flats, hoops, etc. may be produced by the use of rollers . . . with grooves and collars as required".

In his second patent, after describing his method of making from cast iron a fined or wrought iron by puddling in a coal-fired reverberatory furnace, he shingled the product under the forge hammer to slabs "to the size of the grooves in my rollers . . . [and] are worked by me through the grooved rollers" as in his first patent.

The description in these patents of the use of grooved rolls is clear and unambiguous. In the second, the shingling of the bloom to slabs and the finishing to bars in grooved rolls was entirely a novel practice. The first patent was for more than the use of grooved rolls to form bars, bolts and half-rounds for it combined this with improving the grain or toughness of wrought iron by faggoting and reworking. Faggoting was a practice which gave the wrought iron product greater homogeneity and it is well known that "if wrought iron is . . . piled [faggoted], reheated and re-rolled, the physical properties will be improved, and this will continue to about the sixth reworking".[6]

It must therefore be concluded that whilst Henry Cort was not the first to propose, or even to patent, the use of grooved rolls for profile rolling of rounds and other shapes he was the first who clearly established, by records available in contemporary literature, that he could—and did—not only finish blooms of wrought iron to bars but also produce round iron and other shapes by the use of grooved rolls. This method of his, with his puddling process, became the foundation of the rapid development of the British iron industry in the last decade of the eighteenth century and subsequently of iron industries of other countries. His method of using grooved rolls is now the basis of most finishing methods in the steel industry all over the world.

CHAPTER SIX

Henry Cort's puddling process

AS WE saw in Chapter One, many attempts had been made during the first half of the eighteenth century to fine pig iron with the aid of coal in a reverberatory furnace, among them those of the Cranage brothers, the Wood brothers (potting and stamping) and Peter Onions. It is now possible to evaluate Cort's second innovation in the light of these developments.

Having succeeded in making new hoop iron from old with the assistance of rolling techniques, Cort was encouraged to make wrought iron from ballast, a crude form of cast iron used in the navy. This he succeeded in doing and his patent No. 1420 of 12th June 1784 was the result.[1] For this purpose he used a reverberatory furnace, heated by raw pit coal, and to its dished hearth he either conveyed molten metal in ladles or put in the pig or cast iron and closed the charging door until "the metal is sufficiently fused", when the workman

> opens a small aperture ... in the bottom of the doors ... and then the whole is worked and moved about ... by means of iron bars ... in such a manner as may be requisite during the remainder of the process. After the metal has been for some time in a dissolved state, an ebullition ... takes place, during the continuance of which a bluish flame ... is emitted; and during the remainder of the process the operation is continued (as occasion may require) of taking, separating, stirring and spreading the whole about the furnace till it loses its fusibility and is flourished or brought into nature ... and the whole of the above part of my method ... is substituted, instead of the use of [the] finery ... [The second part was] ... to continue the loops in the same furnace ... and to heat them to a white or welding heat and then to shingle them under a forge hammer ... [to slabs] to the sizes of the grooves in my rollers, through which they are intended to be passed, in the manner which I use bar or wrought iron faggotted and heated to a welding heat for that purpose.

He drew attention to the fact that after the first part the iron could be "stamped into plates, and piled or broke and worked in an air furnace, either by means of pots or by piling ... in any of the methods ever used in the manufacture of iron from coke fineries without pots" but that his method could be completed without using "finery, charcoal, cokes, chaffery or hollow fire, and without requireing any blast by bellows or cillinders . . . or the use of fluxes".

This is the clearest account of a process of fining cast iron to wrought iron in any of the patents previously quoted. It will be observed that the term *puddling* which was later applied to Cort's process was not used. It was applied to his process during the last decade of the eighteenth century for, when Cort's patents were in effect cancelled in 1789, his patent or his process of fining could no longer be referred to conveniently and the term 'puddling' was applied to it and to the furnace. Puddling was a word used by the clay workers. The Oxford Dictionary gives for the verb 'to knead and temper a mixture of wet clay and sand so as to form puddle' (1762) and, as applied to iron manufacture, 'to stir about and turn over molten[2] iron in a reverberatory furnace, so as to expel the carbon and convert it into malleable iron' (1798).

Although in a charcoal finery the finer used a rod to raise the iron lumps in the hearth to the oxidising blast and finally to gather and knead the particles into a ball, such stirring was not equivalent to that in an air furnace where the oxidising atmosphere was drawn over the materials in the hearth to the chimney. Instead of—as in a charcoal finery hearth—lifting lumps from the lower levels of the hearth to the blast pipe (which was surrounded by charcoal) the puddler working with an air furnace stirred to expose iron to the oxidising atmosphere (which was unshielded by fuel) and the more oxidisable silicon was oxidised to silica. The carbon then burned with the blue flame of carbon monoxide; some iron was oxidised to ferrous oxide and finally the phosphorus was oxidised, the silica and the phosphorus oxide reacting with the iron oxide to give slag or cinder. In a charcoal finery the particles were lifted; in Cort's air furnace they were raked, stirred and spread. It may be added that it is difficult to see how Onions's furnace could work, for putting the blast below the fireplace heaped with coal could not give an oxidising atmosphere and his air pipe above the hearth was supposed to be 'for occasional use'. It is likely, as the previously quoted extract suggests, that Onions later tried different furnaces not described in his patent.

The hole through which the puddler stirred the contents of the hearth was at right-angles to the flow, from the fireplace to the chimney, of hot gases which heated the arch of the furnace and radiated heat onto the hearth. The stream of air which would enter through this hole would be below the column of hot gases passing to the chimney. This and the molten iron oxide protected the iron from the sulphur in the hot gases from the coal. In 'coke fineries with stamping' which were widely used at ironworks with coke blast furnaces when Cort introduced his patent, the melting of the pig iron in a coke bed removed silicon and converted the pig iron from grey into white iron. This, after being cast outside the furnace and stamped to small pieces, was heated in pots in an air furnace and could not be stirred. The stirring—'puddling'—in Cort's air furnace was unique as the chief process applied to pig iron melted in the same furnace, though there were minor processes for working up scraps in an air furnace which were later quoted as anticipating Cort's chief patent.

The fact that the chief reaction was the oxidation of the carbon in pig iron could not be stated at the time, since the significance of carbon in iron and steel, the composition of air, the part the oxygen of the air takes in combustion and the relationship between metals and their oxides was not made clear until Antoine Lavoisier published his *Traité élémentaire de chimie* in 1789, which was translated and published in Edinburgh in 1790, and received with scepticism by more conservative chemists.

An invention is one thing; its adoption as an innovation another. It remains to describe the development of puddling and rolling at Fontley and the gradual spread of the processes to other ironworks in different parts of the country.

CHAPTER SEVEN

Testing Cort's iron in the Naval Dockyards 1783–6

IN 1784 Henry Cort gave demonstrations of his finery process for making wrought iron in Scotland, Staffordshire and Shropshire and of his rolling process in a rolling and slitting mill on the river Stour. At his instance, in the period 1783–6, following exhibitions of his processes at Fontley Mill, tests were made in the royal dockyards of the quality of iron made under both his patents.[1–3] Tests were made at Portsmouth on mooring chains; and at Portsmouth, Deptford, Woolwich, Sheerness, Chatham and Plymouth on the use of iron for anchors, hooks and ships' bolts, comparisons being made in all tests with similar articles made from Swedish Oregrund iron.

These tests were very thorough and mark one of the earliest comparisons of the relative values of two types of iron (a practice now commonplace) using standard tests. The Oregrund iron used as a reference was made from ore from the famous Dannemora mine, 30 miles north of Uppsala, Sweden, smelted into a white cast iron in a charcoal furnace and then fined to wrought iron. This reference iron, now known to be very low in sulphur and phosphorus, was then said to be "whiter than common iron, and is less liable to rust, is distinctly fibrous in its texture, and is much stouter than any other iron".[4] The production (known as Oregrund from the Baltic port from which it was shipped) was 4000 tons p.a., the whole being later sent to England to Messrs Sykes of Hull. Much of it was used in Sheffield for making cementation steel (and Huntsman's cast steel). The best mark then sold at £40 per ton and another at £39, when the best Russian mark seldom fetched a higher price than £20 per ton.[4]

It will be recalled that in September 1782 Messrs Cort and Jellicoe, having established that they could supply mast hoops of suitable quality,

7 Testing Cort's iron in Naval Dockyards

had had their names inserted into the contract previously held by Morgan for supplying mooring chains and other naval ironmongery to Portsmouth. In 1783 Cort and Jellicoe requested the Navy Board in London to give instructions for the inspection of their new process for making mast hoops, etc., in accordance with the first patent of 1783. Such instructions were given:[1]

> Navy Office, 13 June 1783 to Messrs Cort and Jellicoe
>
> In return to your letter of yesterday requesting that directions may be given to inspect the nature of your new-invented method of working iron, acquaint you that we have directed some of Portsmouth officers with the master-smith to proceed to the works and report fully their opinion.
>
> Chas. Middleton, J. Williams, Geo. Rogers

On 7th July 1783 a report was made[1] to the Navy Board by the master shipwright and his assistant, the master-smith and the storekeeper who "had been to Messrs. Cort and Jellicoe's mills at Fontley, and thoroughly examined the process they make use of in preparing, welding, and working various sorts of iron". This process consisted in putting the iron into an air furnace until it was between a welding heat and a fluid state, when it was either worked under a forge hammer of 7 or 8 cwt or passed between two rollers each of about 9 cwt which "immediately and very visibly destroys the greatest part of the dross and impurities ... and if required will also weld two bars together at the same time [and] bring the most brittle iron into a malleable and tough state ... equal to the best Oregrund iron".

The observers put a 9 lb piece of the best Oregrund iron into the air furnace and when it had acquired the required state put it through the rolls; they recovered a weight of 8¾ lb and concluded that ¼ lb was "dross". Similar tests on "a second sort of iron" lost 2¼ lb from the original weight of 9 lb and "a very bad iron" lost more than half its weight. They deduced that the process eliminated the dross and that the residues were "equally pure to any iron that can be manufactured".[11]

They also tried Cort's faggoting process on old mast hoops; they put the faggot under the forge hammer and, after reheating, passed the product through the rolls and slit some into mast hoop iron. Another similar test gave a final product of ½ in. square bar which drew into bolt-staves "which was very good". By the same process they made a tackle hook and, when cold, tried to break it under the forge hammer but could not. They also drew two links of large mooring chains from the air furnace under the forge hammer and found them "much tougher than those made from the common forge".

They concluded that the process of Cort's first patent yielded from "old and Birstrel [burned] iron", iron which was "very proper iron for making mooring chains, boltstaves, tackle hooks, or any other uses it may be wanted, for this or any of his Majesty's yards".[1]

The letter was signed by "Geo. White, Wm Rule, J. Badcock and J. Greenway", presumably the two shipwrights, a master-smith and the storekeeper previously mentioned and certainly competent to judge; their opinions were very favourable to Cort's first patented process.

A further comparison was made about a year later when C. Middleton and three others from the Navy Office wrote on 1st March 1784 to the "respective officers at Portsmouth yard" requiring them "to demand from Messrs. Cort and Jellicoe a number of links and shackles from mooring chains made . . . according to their own process of working the iron" for an experiment comparing them with a like number prepared from the best iron. Three months later the report on Cort and Jellicoe's mooring chains from iron made "according to their new process of working the iron" against those made "from the best iron in our smithey" was as follows:

> We are humbly to acquaint you that we have made the experiment on the links and shackles on two anvils, and letting the large hammer of four cwts, which shuts the arms of anchors, fall the height of sixteen feet on the centre of the links and shackles between the anvils . . .[2]

Unfortunately this fragment of the report is all that has survived and merely shows the method of testing and not the result. It seems likely that the 'new process' was still that of the first patent, the novelty on this occasion being a direct comparison with similar links and shackles made from best Swedish iron.

There is, however, no ambiguity in a letter from the Navy Board dated 16th July 1784 signed by C. Middleton, J. Williams and G. Marsh and addressed to Mr. Cort which said:[5]

> We have rec^d y^r L^r of yesterday with a bar of Iron made of Ballast Iron which you inform us can also be made from Shot, Shells etc, and desire you will make your proposal for work^g up ye diff^t sorts of old iron in ye Kings Yds for diff^t uses. We approve much of the Bar & shall treat with you if your terms are reasonable.

The sequel to both these approaches has been recorded:[6]

7 Testing Cort's iron in Naval Dockyards

The respective officers of Portsmouth Dockyard by order of the Navy Board, attended at various experiments made by Mr. Cort's patent methods, and in their reports, the Navy Board were pleased, on 12th July 1784, to contract with Cort & Jellicoe for converting their neutral[7] Iron into mooring chains, by his new improved method, and on 18th August 1784, they converted their cast iron ballast, and old iron, into mooring chains by his improved method; and on 22nd October 1784, they contracted for converting their unsizeable Sweeds[7] iron into suitable sizes by his improved method.

These demonstrations and tests during 1784 were followed by more systematic tests on the P- and eye-bolts for ship's caps, tackle hooks, top-tackle snatch blocks, and on anchors in six different naval yards, starting in 1785. In preparation for these tests 60 tons of old ballast, rusty and dirty, were delivered to Mr Cort. Some of the ballast was in pieces of 5, 6 and even 7 cwt which could not be broken and such pieces had first to be melted into pigs before being put into the 'refinery furnace'. Despite this there was still a very irregularly sized feed to the furnace which was operated by two men, with one of three resting in turn. This team "with ease refine 4 tons of pig a week from one furnace", the product being rolled at one welding heat into bar of the smallest dimensions at the ratio of 20 cwt from 32 cwt of pig iron, though Cort only returned 29.15 tons of bar.[8]

From this bar six large anchors of 34 to 59 cwt were made and comparable anchors were made from Swedish Oregrund iron by the dockyard smiths. They also made a variety of other articles of naval ironmongery from both Cort's test iron and Swedish iron. The method of testing the anchors is detailed in the report from Deptford: the arms of the anchors were interlocked and drawn against each other by 30 men heaving at each of two opposed capstans. The hooks were tested similarly.[9]

Anchors made from Cort's fined iron proved superior in five out of six tests and in the other did not suffer in comparison with the Swedish. Although in the totals for similar tests on hooks, more of Cort's than of Swedish iron broke, fewer of Cort's straightened and more of Cort's survived without damage. The Woolwich results were especially favourable to Cort's hooks, suggesting that there was probably a factor dependent on manufacture. The overall conclusion from Portsmouth Dockyard that articles made from Cort's iron were equal to like articles made of the best sort of Oregrund iron was amply supported in all other dockyards.

Henry Cort: the Great Finer

The series also included tests at Deptford on eye-bolts in which two, of each sort, "appear to have driven equally well". In the same number of tests on P-bolts, Cort's drove without injury; both the Swedish bolts drove but one broke in the stump. At Woolwich it was said that Cort's iron was "rather more malleable, retains its heat longer, welds and bears punching, turning and being wrought in the most severe manner to the full well".

The general comparison with "the first sort of Oregrunds iron" should be appreciated, for there was a considerable differential in the price of this and ordinary Swedish iron. On 1st April 1769 Oregrund Firsts cost £22 2s 6d, Oregrund Seconds £19 12s 6d and Stockholm iron £17 12s 6d per ton delivered at Portsmouth. Earlier, on 2nd June 1765, the first two varieties each cost 7s 6d less per ton.

Dr Joseph Black of Edinburgh, in a letter of 15th May 1786, published in *Brief State of the Facts, 1787*,[10] said:

> I meant also to have explained to you my opinion of Mr. Cort's process of making iron; it is this: I have never analyzed the iron made by that process, as I always considered direct experiments to prove its toughness in its hot and cold state, as also its strength and other good qualities, as the most interesting and decisive trials, and I like the process, and the iron, for the following reasons—

He enumerated four: (1) that it was made using coal; (2) that it was heated by flame and not mixed with the fuel and ashes in the common (charcoal) process; (3) that rolling forced out the melted slag, giving the iron more cohesion than the common method with the hammer. The fourth was:

> Fourthly: By the experiments made here, I saw that Mr. Cort's iron was exceedingly soft and malleable when hot, and very tough when cold; and I have heard of more decisive experiments made in England which prove it to be possessed of very good strength and toughness; for these I refer to Mr. Cort, who, I suppose, can give evidence of them. The only point which remains undecided with me is, whether Mr. Cort's iron can be afforded sufficiently cheap; and this point will be surely decided by any company who may establish works on his plan. I am informed that a great deal of bar iron is now made in England sufficiently cheap, by another process, in which also the expense of charcoal is avoided, but such iron is not of the best kind.

The result of these trials (completed in August 1786 according to Dr Black's letter) was that the Navy Board

> were pleased in April, 1787, to contract with Cort & Jellicoe for 150 tons of Iron, to be made according to the said new process, and in April, 1788, they

7 Testing Cort's iron in Naval Dockyards

were pleased to contract for the like quantity made by the said process which had also been sent to the different Dock Yards, and reports thereon having been made by the several respective officers, the Navy Board were pleased, in April 1789, to advertize for 200 Tons of Square Feather edged Iron, and that no tenders would be regarded but from persons who can prove that they make it agreeable to Cort & Jellicoe's Patent, and in September 1788, they were pleased to contract with Cort & Jellicoe for small ship's bolts to be converted from old iron according to H. Cort's patent method. The making of feather-edged bars by H. Cort's process is of great utility to the forming of shanks of anchors, and is attended with a considerable saving to Government.

We have so far quoted only the description of Henry Cort's fining process given in his second patent. In *Brief State of the Facts, 1787*,[11] there is a full description of the process seen at Fontley by David Hartley, Esq.:

Golden Square, 19th January 1786

Having heard that Mr. Cort had discovered a method of making the very best iron out of common iron ballast, by a short and simple process, I went to his works ... The process, as I saw it three or four times over, is something to this effect: Between two and three cwt. of common ballast is melted in an air furnace with sea-coal. When melted, it spits out in blue sparks the sulphur which is mixed with it. The workman keeps constantly stirring it about, which helps to disengage the sulphureous particles, and, when thus disengaged, they burn away in blue sparks.[12] In about an hour after melting the spitting of these blue sparks begins to abate (the workman stirring all the time), and the melted metal begins to set ... the clotting of the metallic particles by the stirring about may be compared to churning. As the stirring of cream, instead of mixing and uniting the whole together, separates like particles to like, so it is with the iron; what was at first melted comes out of the furnace in clotted lumps, about as soft as welding heat, with metallic parts and dross mixed together, but not incorporated. These lumps when cold resemble great cinders of iron; they are called loops.

The next part of the process is to heat these loops to the hottest welding heat in an air furnace, and then put them under a great forge hammer, which by a few strokes at the very highest point of the welding heat, consolidates the metallic parts into a slab of malleable iron, about three feet and half long and three inches square. The hammer at the same time expels and scatters the unmetallic dross. These slabs are brought to a wedge point at each end. They are malleable iron, but still with a considerable mixture of dross.

The next part of the process is to heat these slabs to the hottest welding heat in an air furnace, and then to pass them through the rollers of a rolling mill; the slabs being extremely soft at the highest point of welding heat, the force of the rollers consolidates the metallic parts into bar iron, and the dross is squeezed out and falls under the rollers. This is the whole process;

Henry Cort: the Great Finer

and thus in about six hours I have seen a piece of common iron ballast rolled out into a ship's bolt; I have seen this bolt laid hollow across the eye of a large forge hammer and receive two hundred and fifty strokes of the heaviest sledge hammer, and thus bent double, but without breaking, or suffering the least apparent injury . . .

It is to be observed likewise that the common blooms, as they are called, in ordinary forges of iron, are nearly three times as thick and solid as the slabs in Mr. Cort's process, and therefore much less affected by the blow of a hammer than the slabs are under the effect of the rollers. His slabs are small, soft, and ductile, and therefore easily suffer the expulsion of the dross by the squeezing of the rollers.

The final extract we can take, in this chapter, from *Brief State of the Facts, 1787* is the well-known statement of Lord Sheffield:

If Mr. Cort's very ingenious and meritorious improvements in the art of making and working iron, the steam engine made by Boulton & Watt, and Lord Dundonald's discovery of making coke at half the present price should all succeed, it is not asserting too much to say that the result will be more advantageous to Great Britain than the possession of the thirteen Colonies, for it will give the complete command of the iron trade to this country with its vast advantages of navigation.

Henry Cort was gratified that Lord Sheffield had spoken so highly of his processes, and wrote to him, referring to the question as to whether the import of foreign iron should be prohibited. He suggested that if a loss of revenue from tax set against the sums paid annually to Russia and Sweden for the purchase of the iron should show a balance, it would be thought worth while for Parliament to repay the money he and his friends had expended in perfecting the process, when he could make a gift of his patents for public use. He went on:

. . . it seems hard that the most generous Part of the Trade should pay me for the use of my Inventions, whilst others should have the same advantage, and pay nothing for it. This makes me desire to put all the Trade on a footing by accepting from Government (if I could obtain it) that Reward which I may have much difficulty to procure from Private Persons.

This statement was to be prophetic of the difficulties to come.

CHAPTER EIGHT

The adoption of Cort's processes (1) Coalbrookdale

BY 1784 Henry Cort was demonstrating his processes to other ironmasters, among whom were the 'Coalbrookdale Group'. In the period of negotiation of Henry Cort with the Coalbrookdale Group from 1784 to 1789, Abraham Darby III was still associated with the group—Coalbrookdale, Horsehay and Ketley ironworks—but the negotiations were concerned with developments at Ketley where the Reynolds interest was strongest. Richard Reynolds had in 1775 increased his share in Ketley and Horsehay works from one-third to two-thirds by acquiring the Goldney shares, had advanced £20,000 to Abraham Darby III to purchase other Goldney Coalbrookdale shares and in 1780 became the lord of the manor of Madeley and the landlord of the Coalbrookdale company. In 1784 he started the Donnington Wood works and his son William became interested in Cort's work to enable a decision to be made on the form of the fineries to be installed there, as well as in the developments proposed at Ketley. By the end of the period we are considering Richard Reynolds was transferring his ironworks shares to his sons, Abraham Darby III was dead, and we are concerned with the negotiations conducted by William Reynolds with Henry Cort. It should be remembered that the Coalbrookdale Group was, in this period, still the most important ironmaking group in Great Britain and only John Wilkinson's scattered interests in three counties were comparable in scale.

The Weale MS records the early correspondence from William Reynolds of the Ketley company to Henry Cort, and indicates the state of progress of rival processes:[1]

17th January 1784

I am much chagrined that Peter Onions has not yet been able to succeed in his furnace, he is going to try in one built some time since by the Dale Company which I hope will prove more propitious to his wishes.

Wm Reynolds

Henry Cort: the Great Finer

17th February 1784

To Hy. Cort, Gosport

Respected Friend,

 I am very sorry to say Peter Onions has not succeeded & has with us entirely given up the point, however he seems confident of succeeding elsewhere, but this I doubt, unless he hits upon some other expedient. I am on the contrary as much pleased to hear of the great success of thy experiments, and sincerely hope they will be attended with every advantage to the ingenious author he can wish and am

   Yrs
   Wm Reynolds

The Weale MS continues:

In the course of the year 1784 Mr. Reynolds erected very large new works at Ketley where he has laid out many thousands of pounds all of which were constructed . . . with a view to work according to the method commonly practised viz by Fineries blown by Cylinders with the assistance of Steam Engines but hearing that Mr. Cort intended to exhibit his process publicly in Staffordshire, Mr. Reynolds delayed building his Fineries in order to see and judge of Mr. Cort's process before he completed his plan. Accordingly, Mr. Cort exhibited his process in the neighbourhood of the Dale and Mr. Reynolds having been present and seen it and being convinced of the advantage of the process has changed his plans and instead of proceeding according to the old method he intends to adopt Mr. Cort's—many other Ironmasters had previously adopted the process and agreed to give Henry Cort 10/- per ton for the privilege of using his method.[2]

On 6th December 1784 there was a special demonstration of Cort's finery process at Ketley ironworks and Cort took the precaution to have this minuted:

              Ketley 18 December 1784

Mr. Reynolds saw H. Cort make tough iron and coldshort iron from the same materials at one heat, by a variation of the process

   Wm Reynolds

There is a letter which gives a report on this trial:

The half-bloom made at Ketley forge by Mr. Cort's men from the Ketley coak pig iron has been drawn out into bars by us at the lower forge in Colebrook Dale. The iron bears an exceeding good heat, works well, and

8 Adoption of Cort's processes: Coalbrookdale

soft like wax under the forge hammer, and drawn into good merchantable iron

> Dated 22nd Jany 1785 at Colebrook Dale,
> Thos. Cranage, Thos. Jones
> Hammermen to Messrs Reynolds & Co at Colebrook Dale

A further comment is of interest:

> The Dale Co. have said that the reason why they left off Cranages and adopted the granulating method was that the iron produced by the former was not fit for nails the making of which is the greatest consumption for iron in Staffordshire and Shropshire—at the time they made the declaration it was supposed that H. Cort's iron was not fit for nails but it has since been proved to be very fit for the purpose.[3]

The following letter proves it:

> Joseph Amphlett to Henry Cort   Dudley, 17th January 1785
>
> Dear Sir,
>
>   Mr. Homfray sent me three bundles of iron made from his Welch pig by your process at Wednesbury all of which I have worked and for your satisfaction I will give the just opinion of the Nailors that worked them. [The nailors said that it was either 'good iron' or 'better iron' and that they got a good yield.] From the opinion of the three different workmen and from the nails which were tough and good I should be glad to buy a considerable quantity of such goods as Mr. Homfray's Welch pig produces.

There was a further letter which refers to nail making, apparently from the wrought iron bar made from Ketley pig which had been rolled and slit to nail rod at West Bromwich:

>   26th January, 1785
>
> Respected friend,
>
>   Foxall has brought back from West Bromwich a bundle of the rods drawn out there which look exceedingly well and which our man says bears the rolls and cutters as well as any iron could be expected to do. I intend dividing them, and sending them to different nailors for trial
>
>   I remain, with much respect, thy assured friend
>   Wm Reynolds

These records show that the demonstrations which Henry Cort had arranged at Shut End (Staffordshire) in June and at Ketley in December 1784 had given convincing evidence of the quality of the wrought iron produced both from Ketley and 'Welch' coke pig.

The evidence suggests that Cort claimed he could make a ton of bar iron with 30 cwt of cast iron; Reynolds and Co. took thirty-two hundredweight to make a ton, and other statements suggest that the stamping method was being used at Ketley.

Despite the above statements concerning the trials at Ketley, the Coalbrookdale Co. refused to pay Cort a royalty, alleging similarities between his process and that previously practised by the Cranage brothers at Coalbrookdale.

The information in the Weale MS is gathered in such a way as to suggest that Henry Cort had considered testing the validity of his patents in the courts. Part of a lawyer's comment runs thus:

> ... It is submitted upon the whole matter that Mr. Cort's discovery, as far as it is beneficial, is entirely new & that the ineffectual attempts which have been made in the use of air furnaces before with the same view, do not detract from the novelty of the invention which is effective and as to this the testimony arising out of the conduct of the Coalbrook Dale Co. is peculiarly strange, who after giving the best of former attempts an ample trial without success are going to adopt it as their mode of working.

This was confirmed by the opinion of Alan Chambré (1739–1823), then a barrister at Gray's Inn:

> It seems to me that if Mr. Cort determined to enforce his patent, and to bring actions for infringements upon it, he had better contend with that company (Coalbrookdale) than with any others, for their conduct affords many circumstances which could not so well be given in evidence in a contest with others, and which may be very material; and besides, if Mr. Cort proceeds against persons setting up a rival patent, and pretending the merit of first discoverers, it is not likely that others will hereafter contest the point with him.
>
> A. Chambré, Gray's Inn, 4th March 1785

It appears that the advice in these two opinions was not acted upon. This missed opportunity may have been the reason why Cort, in his letter to Lord Sheffield three years later, talked about a grant from Parliament. After 1789 no action for infringement was possible because the patents—as will be seen later—were impounded by the Navy. So passed all possibility of establishing the validity of his chief patent when it had only run five of its 14 years. In the next chapter we shall record Henry Cort's dealings with Richard Crawshay of Cyfarthfa where the greatest developments of Cort's processes occurred during the legal period of his patents.

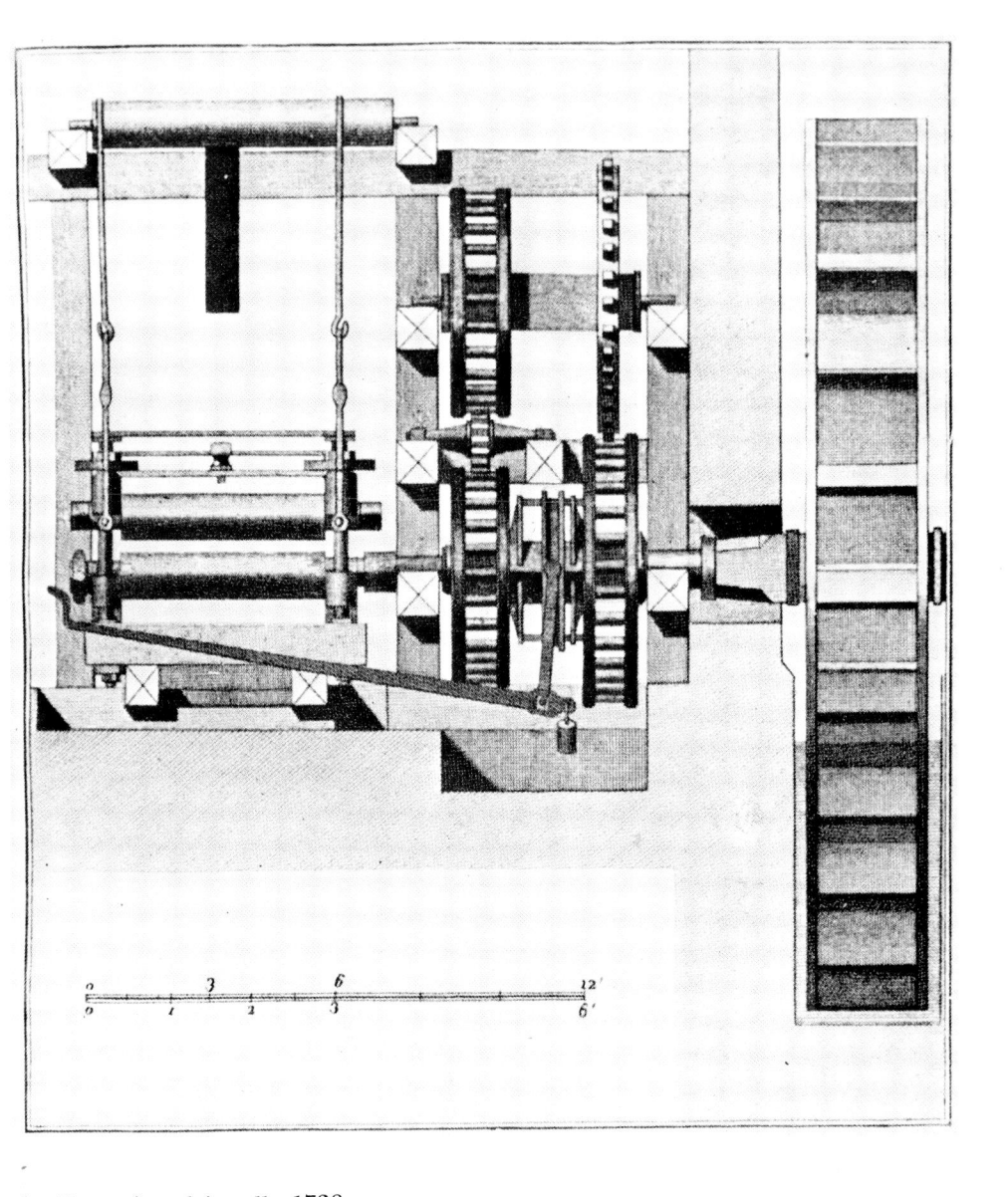

1 **Reversing plain rolls, 1728**
(from 'Encyclopédie Méthodique', T. 3, 1728, Pl. 12 as reproduced in Johannsen)

2 Rolls in a slitting mill
(after Swedenborg, 1734, 'De Ferro', Tab. 29)

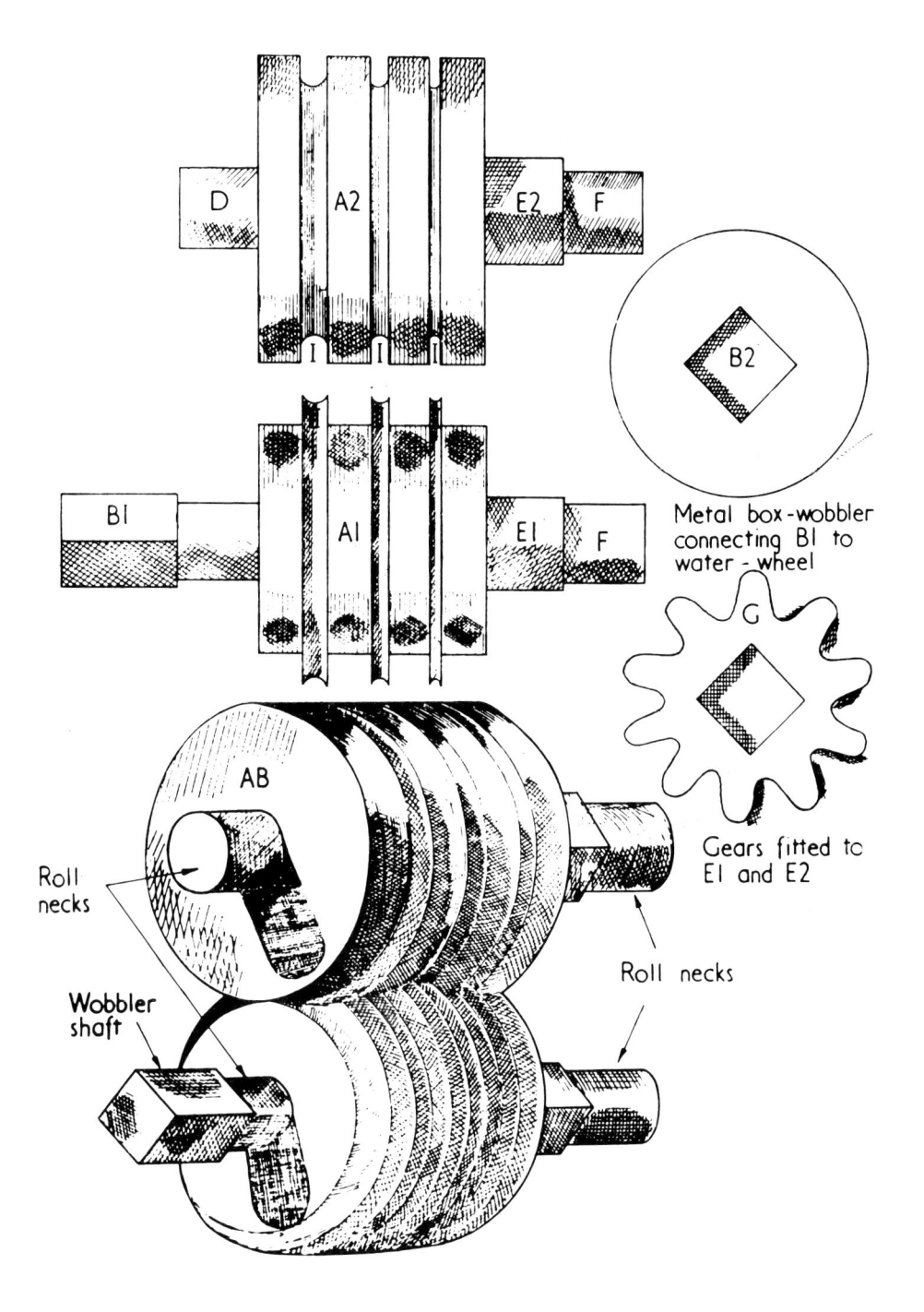

3 Grooved rolls for rounds, 1766
(from Purnell's Patent)

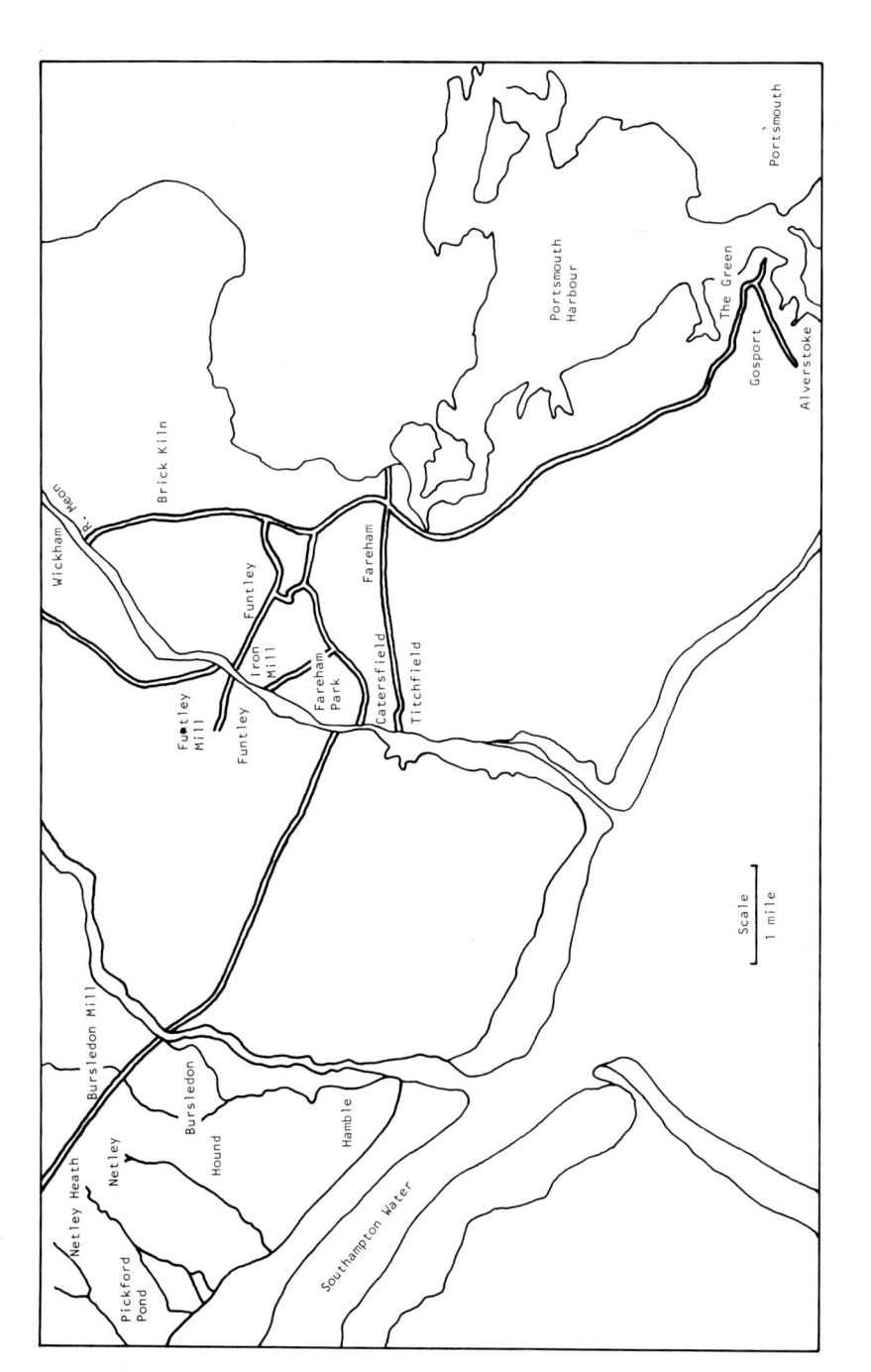

4 **Fontley iron mill in relation to Burlesden furnace and Gosport smiths' shops, based on the 1810 Ordnance Survey map**
The spelling of Fontley is variable; while the O.S. (as shown here) used u, it is believed that at the time of Cort it was spelt with an o

5 Tombstone of Henry Cort, Hampstead 1966

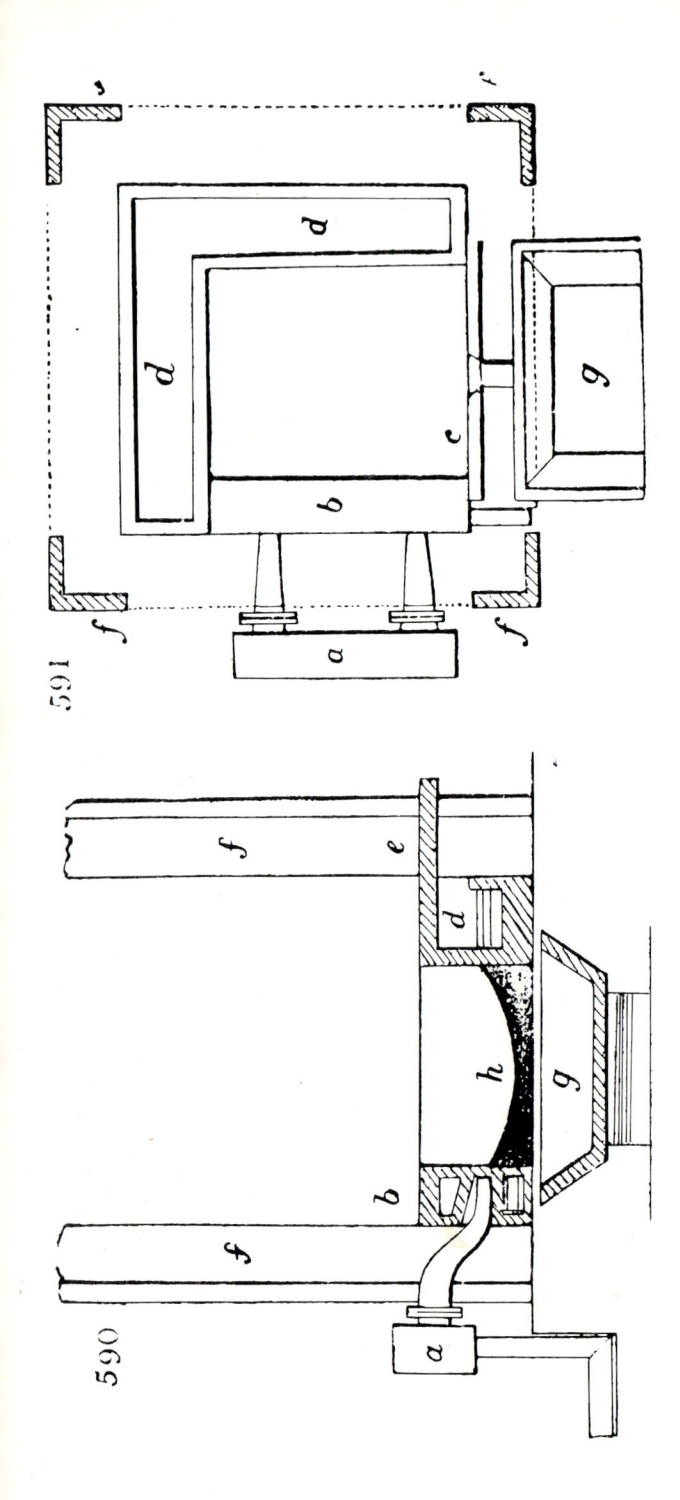

6 **Coke refinery, Low Moor**
(This and Plates 7 and 8 from
'Dictionary of Arts, Manufactures and Mines' by Andrew Ure, first edition 1839)

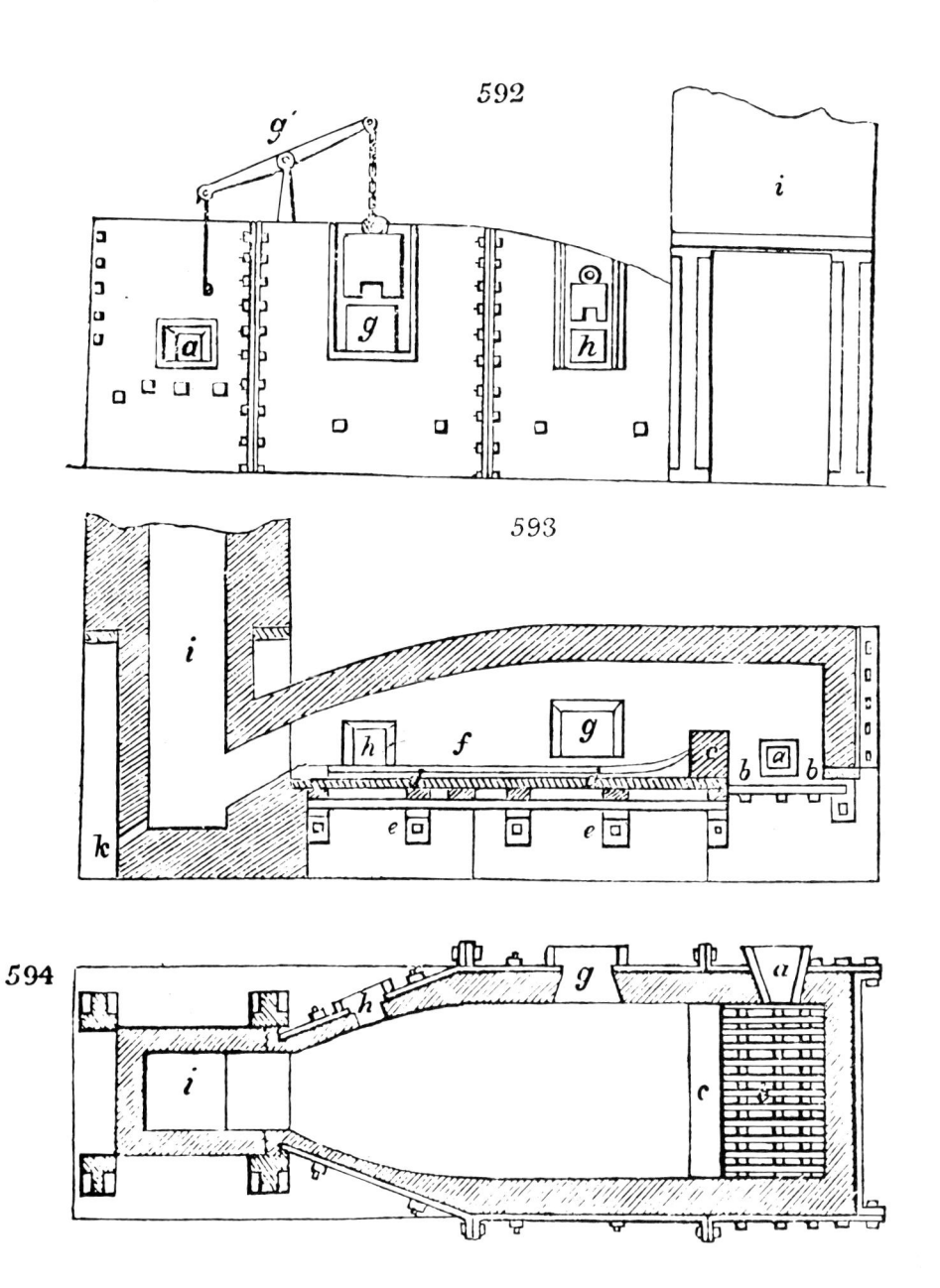

7 Puddling furnace, circa 1830
(Ure)

8 Plate rolling and shingling mill with elliptical and rectangular grooves, circa 1830 (Ure)

CHAPTER NINE

The adoption of Cort's processes (2) Developments at Cyfarthfa 1787-9

THE FIRST contact of the Welsh ironmakers with Henry Cort was in June 1784 when Francis Homfray of Wollaston Hall, Worcester, arranged a trial of Cort's rolling process at the rolling and slitting mill at Hyde, Stourton. Homfray had, from 1782 to 1784, leased the gun-casting foundry and forge of Anthony Bacon at Cyfarthfa, Merthyr Tydfil, Glamorgan, and was supplied with pig iron from one of the furnaces at Cyfarthfa or Plymouth (which Bacon owned) or Hirwaun (which Bacon at first leased and finally owned). Since the end of the American war was in sight when Homfray took over the gun foundry it is not surprising that, on the diminution of orders when it ended, he surrendered the lease in October 1784. Nevertheless, he had appreciated the potentialities of the South Wales coalfield and further developments of Homfray interests were to occur.

In 1784 Jeremiah and Samuel, the eldest and youngest sons of Francis Homfray, obtained a lease of an ironstone field in Merthyr Tydfil for 99 years. They then acquired sufficient surface properties at Penydarren, within the parish of Merthyr, for a blast furnace to be built there in 1785. In June 1786 the brothers received a lease of a coalfield from the Dowlais company, near Pyllyhead, to begin in 1792 but had apparently also acquired sufficient coal to start their furnace forthwith. The test of Cort's rolling process in June 1784 was probably part of their general enquiries but, in fact, they built their first forge after the pattern at Cyfarthfa, i.e. to use 'stamping and potting'.

Richard Crawshay had held a one-third interest in Anthony Bacon's gun-casting foundry since 1774 but when Bacon died in January 1786 and his estate went into Chancery, Crawshay acquired the lease of Cyfarthfa furnace, foundry and forge on 12th February 1787. He was partnered by

51

William Stevens and James Cockshutt, both of whom had been partners in the gun foundry. Cyfarthfa, it will be recalled, had been built in 1766 by Charles Wood to use his 'pot' process of making bar iron, so that there was a longer experience of this at Cyfarthfa than elsewhere. The Board of Ordnance contracts for guns must have been very small after the American war and the making of wrought iron would have been dominant in the thoughts of the partners at Cyfarthfa.

The first date in the Weale MS about the association of Henry Cort with Cyfarthfa is 1st May 1787, when an agreement to pay a royalty on iron made according to Cort's patents was signed, but this must have been preceded by tests on Cyfarthfa pig iron.[1] Richard Crawshay who, when he signed the agreement with Cort, had obtained the lease of Cyfarthfa furnace less than three months earlier, was aged 47. He was a newcomer to Wales and to ironmaking, for he was an ironmonger of London whose interest in gunfounding had been financial. The contract for the supply of pig iron from Plymouth furnace to Cyfarthfa foundry and forge, made by Bacon and Homfray in 1782, continued until 1793 (since the Plymouth estate was in Chancery) and was at the rate of only £4 per ton, probably one of the cheaper pig irons in the United Kingdom. Crawshay's rent for Cyfarthfa was £1,000 p.a. but he paid no royalty for this coal and ironstone, of which there were enormous reserves. Clearly, he had a great incentive to develop his production of iron, cast or wrought, and the demand for cast iron for ordnance was then limited by peacetime conditions.

Before the agreement to work Cort's processes was made it was proposed that Crawshay should pay a royalty of 15s 0d per ton, the same as had been agreed by the Rotherhithe company,[2] but Crawshay "objected & said that he would never think of payment on the same basis as with the Rotherhithe Co & if he went into it Reynolds and Homfray would do the same". Accordingly an agreement was made for a royalty of 10s 0d per ton. There are several letters of May 1787 minuted in which the correspondents are usually indicated by initials: H.C. (Henry Cort); A. J. (Adam Jellicoe); S.J. (Samuel Jellicoe); and R.C. or Mr. C---y (Richard Crawshay) which it is interesting to record:

9th May 1787: R.C. to H.C. re a shortened version of the agreement
I have no desire of depriving you of the bounty agreed for leave to work on the principles of your Patents directly or indirectly but to pay 10s 0d per ton for all the iron made at my works in that way as proposed here and is now inserted in the altered agreement.

9 Adoption of Cort's processes: Cyfarthfa

11th May 1787: conversation R.C. with A.J.
I give you my word of honour I mean honestly and fairly if I approve of Mr. Cort's plan I will go into it most liberally, but I will not have my hands tied behind me. Do not let us quarrel at the outset I have seen nothing of it yet. If I go into it W. Reynolds and Homfray will follow for I have Iron from them. Wilkinson I know will not but you'll by this means have three of the greatest Houses in the Kingdom.

May 1787: A.J. to S.J.[3]
I have your letter with the agreeable account of what passed on Monday with C---y at Fontley & I hope for a happy conclusion of the business & flatter myself it will prove the providential [event?] of much good. I do not pretend to have much insight into People but if Mr. Crawshay is not an honest well meaning man and does not prove himself so I shall be very greatly mistaken.

22nd May 1787: A.J. to S.J.[3]
He [R. C---y] told me Mr. Reynolds was in a very harmful situation being quite at the mercy of his workmen that this process would set him free and he thought he might by degrees overcome the prejudice of the Ironmasters. Cort more cool and less agitated, for he thought it might hurt his health & spoke of him in very friendly terms as a very honest man & said 'I must take him to Wales & he must be my pupil altho' he is my Master in Iron Matters'. In regard to what he said of Mr. Cort's being very earnest & agitated on the Business I told him it was no wonder, that he had gone through great trials & that I hoped his anxiety would soon be over, as he Mr. C---y had taken the matter in hand. He agreed to it all & hoped he would soon be calm and easy. I shall ever think it a great providence that Mr. C happened to meet him; this I think the course of our conversation.

Crawshay's second visit to Fontley is recorded:[4]

10 June 1787
Mr. Crawshay and his partner Mr. Cockshutt visited Fontley works and made further trials to their satisfaction. An agreement took place between Mr. C---y and Cort and Jellicoe to make iron blooms for them at his works in Wales. At the request of Mr. C---y, H.C. sent proper workmen to erect Furnaces at Cyfarthfa and to instruct others in the art of making iron by H.C.'s process and that H.C. with his son William at the request of Mr. C---y went there twice to supervise the work carrying out there.

In a letter from James Cockshutt to Coningsby Cort some 25 years later:

In consequence of my examination at Fontley (in 1787), with the assistance of Mr. Cort and several of his workmen the process was begun at Cyfarthfa and a very powerful Rolling Mill erected, and under the encouragement of Mr. Crawshay, who took upon himself the sale of the whole, and therefore was the best judge of its reception among his numerous customers, that the

Iron thus made, was shortly increased to double the quantity of that had been produced by the former method.[5]

On 21st June 1787, Richard Crawshay to Messrs Cort and Jellicoe:

> They are making nearly 15 tons per week in the manner all intended for you.[6]

On 14th September 1787, Mr. C---y to H.C.:

> At the latter end of September 1787 the vessel with the Welch Iron arrived.[7] Mr. Crawshay in consequence paid another visit to Fontley works & there saw the Blooms [that had been made by H.C.'s patent in Wales] rolled into Merchant Bars or half-flat iron of 2 Inch by 1 inch and ⅜th thick, a size never before made in England or any other country from Coak Pigs. Mr. Crawshay on seeing this performed expressed great satisfaction and in the mill in the presence of the workmen laid his hand upon H.C.'s shoulder and said 'My dear Cort, this will do. We shall now make Swedish Iron for sure in England which heretofore we were obliged to have from Sweden' & that he would not have missed seeing the tryal for £500.[7]
>
> The bars were smoothed up under a light hammer & allowed to be as handsome Iron as could be seen in the London Market & are formed more true by the rollers than possibly can be by any hammers.

By October 1787, Henry Cort was again in Wales:

> *25th October 1787, H.C. to Mr. C---y*[9]
> I called on Tuesday at Penydarron Forge with Mr. Cockshutt the hammers work well and saw some very good iron but it happened as we went in that the Bar Iron then drawing dropped in pieces at the red short heat and there had been drawn that morning some others of that quality. We asked from whose pig it had been made and Mr. Homfray said it was Plymouth Pig made in the stamped way—he had worked a few tons but would work no more, that he would not give anythink for them, that he had been informed Mr. Crawshay had contracted for a large quantity on moderate terms (a 1000 tons per Anum at £4). We brought with us a piece of the Bar and had it tried but it was so rotten it could not be worked into anything. He said he would write to you about it & which might be unpleasantly received. I was therefore determined to have some Plymouth Pigs brought from the same heap Mr. Homfray received his. I have worked them by my process better Iron cannot be made, very tough, a strong body intirely free from red short quality, which you will have confirmed from this house, when they have occasion to write to you. Mr. Homfray acknowledged he never expected my process brought to what it is. Mr. Homfray said he could make his Iron by my method forty shillings per ton cheaper than by the method he now pursued but whilst others kept on stamping he should—if others altered their method he should do so too and at the same time was ungenerous to say he would not pay me any thing if he went into it. What I felt is not to be

9 Adoption of Cort's processes: Cyfarthfa

expressed—however my ardour is not to be stated. I am connected with you & I trust there is a Providence yet to support me & mine though an ungrateful set I have to encounter with. I beg my compliments to Mrs. C., & your son.

Crawshay replied on 3rd November:

> I am much pleased at your journey and the three men to teach the Welsh your mode of Making Iron which I shall not repent paying 10s 0d a ton if others reject it . . . Sam Homfray's failure with Plymouth Pig & your succeeding may (it ought to) teach him better manners to you. Your feelings have been so often tried & depressed that as a friend I advise you not to feel sore.

By March 1789 the Homfrays were experimenting with Cort's processes:

> *11th March 1789: Jeremiah Homfray, apparently to Henry Cort*
> Dear Sir, Am much obliged for yours of the 19th February, and the other side send you an account of our performance . . . So satisfied myself about it that I shall provide no more blast than is necessary for blowing a furnace in a new concern I am about establishing upon these mountains. The new forge at Cyfarthfa, upon your plan, is a noble thing, and does credit to those gentlemen there; they are driving it forward with great spirit; and will constantly turn out about 30 tons a week of blooms; they are now drawing some for London. Your rolling mills . . . makes them beautiful iron . . . The Port iron will be slit at Stourton, and worked under my father and brothers Sam and Tom's own eye, who are both very anxious for the well doing, and impatient for its trial.

Then followed details referring to a test on 21st February 1789 in which it was stated that 35½ cwt of pigs made 20 cwt of bars.

Further correspondence in the Weale MS suggests that the successes at Cyfarthfa were attracting the attention of other ironmasters, including Mr Gibbons at his works near Stourbridge, who was considering entering into an agreement to pay Cort 10s 0d per ton.

> *28th June 1788: Mr. Cockshutt to Mr. C---y*
> Since my last have had Sam[l] Thomas Mr. Reynolds Agent at Horsehay. Mr. Reynolds says in consequence of some conversation he has lately had with you & requested he might see our process under Mr. Cort's patent in which we did what we believe you would have done, we used no reserve with him, he saw the whole as well as our state of yields made—he went away very well satisfied and says he does not doubt it becoming the general practice.

55

Henry Cort: the Great Finer

There was further progress at Cyfarthfa:

> *5th July 1788: Mr. Crawshay to Cort and Jellicoe*
> We now make about 20 tons weekly from 6 Furnaces better Iron than ever. Messrs Gibbons & Co want to work on Mr. Cort's plan, if he could step into the breach and come here for a day or two in the middle of the next week, perhaps we might digest the business better for his interest . . .

Nevertheless, there was still some reluctance to agree to Cort's terms, and Crawshay advised Cort on 25th July 1788 to take 5s 0d a ton instead of 10s 0d. Cort agreed, and Crawshay circulated other ironmasters accordingly. For some ironmasters there may have been little cost advantage at this stage in adopting Cort's processes.[10]

> *21st December 1788: John Cooke to H. Cort from Kilnhurst forge, near Doncaster*
> I am informed that, at Cyfarthfa, one Forge is employed in your method, *the other in the Shropshire mode of Fineries* with Coak and Stamping and Balling as it is there called and they cannot decide on the advantage of one over the other—at Derwent Coat[11] near Newcastle, I have been told, the practice is to make Iron in the Test Furnace, but not to ball it therein, but to take it out by small pieces and stamp the cinder out, and when cold to pile the thin pieces in another Air Furnace.

Nevertheless, John Cooke agreed to pay 5s 0d per ton.

The last letter from R. C---y in these records was dated 19th May 1789:

> I received both your letters of yesterday & the rods for coach very much improved since Wednesday, it works well notwithstanding its dusty colour which is a heavy objection here—we will overcome all the difficulties by time and perseverence. I intend going cooler to work than you do—but I will never desert our plan as laid down at your house as long as yourself say 'tis tenable. I have hitherto overcome all difficulties that I encountered & with God's blessing & your assistance I will conquer this—therefore as a real friend I advise you not either to despond or to elate—give Mrs. Cort a little more of your company in the morning—think & talk more of other things than Iron frequently from which you will certainly ease the body and strengthen your mind—both which of late have been severely tried
>
> Your sincere friend & obed[t] Ser[t], Richard Crawshay

On 1st September 1789 "Mr C---y began to work his new Mill upon H.C.'s plan", in a month which was tragic for Cort for in it his patents were seized by the Navy and Cort was served with a Writ of Extent for debt owing to the King of £27,500, which forced Cort to become bankrupt.

CHAPTER TEN

Bankruptcy

IN THE preceding chapter the detailed story was taken to September 1789 when Richard Crawshay brought into operation his new rolling mill with its battery of puddling furnaces, both built to Henry Cort's design and under his supervision. This should have been the turning point for the better in the affairs of Henry Cort, the beginning of a wider adoption of his processes by the chief coke-iron makers—with the payment to him of a royalty of 5s 0d per ton on every ton of bar iron made. In the event it proved the beginning of the disaster which left him bankrupt, dependent on poor relief for five years and to die, after a further six years in lodgings, broken-hearted.

The immediate cause of Henry Cort's bankruptcy was the death of Adam Jellicoe on 30th August 1789.

As we have seen, Adam Jellicoe had had a long association with Cort, probably dating back to 1780 when the former was in the Pay Branch of the Navy, the chief 'out-station' for the payment of seamen's wages being Portsmouth. Jellicoe served in all for 49 years, entering the Pay Branch in 1740, and from 1777 was styled Deputy Paymaster.[1] We have noted that Adam Jellicoe was able to lend Cort money to modify his mill in 1780 and subsequently to advance further sums against Cort's patents as security, a partnership for his son, Samuel Jellicoe, and 50% of the profits of Cort & Jellicoe, with 5% interest on the loan. Henry Dundas, Lord Melville, First Lord of the Admiralty, told the Lords of the Treasury when, with the fall of Pitt's ministry, he left his office in May 1800 that on entering it he had found: "At the Head of the principals of this department Mr. Adam Jellicoe a public Servant, who had successively filled every Department of the Office, and who had risen to the First Situation in responsibility with the most honourable character. His length of service, then about Forty-two years, entitled him to every confidence".

One of Jellicoe's junior clerks was Alexander Trotter (1755–1842) who had entered the Navy Pay Office in 1776 at a salary of £50 p.a.[2] and served in it until 1784 (by which time his salary was under £100 p.a.). He was out of the Pay Office for a year and in December 1785, on the death of Andrew Douglas, returned as Paymaster of the Navy at £500 p.a., thus becoming Adam Jellicoe's superior at the age of 30, and at a time when a new Act (25 Geo. III, c. 31) of 1785 was coming into effect to regulate procedure in the Office of the Treasurer of the Navy.

Although the new Act directed that all issues to the Treasurer of the Navy should be placed to his credit at the Bank of England, thence to be drawn by drafts specifying the heads of service for which they were wanted, this requirement, although it discouraged, was not sufficient to prevent the use of balances for private purposes and Trotter found a means of circumventing it. He told the Commissioners of Naval Enquiry[3] that:

> it is no doubt generally presumed, that advantages are derived from the use of public money by those in whose hands it lies. I have myself no doubt, though I cannot prove it, that such advantages were enjoyed by my predecessors . . . and the exceeding smallness of the salary of the paymaster, when compared with the immense extent of the trust and responsibility reposed in him . . . afford a presumption that such advantages have been considered as forming a part of the remuneration of so anxious and confidential a charge.

During his period as Paymaster, a total of £134 million[4] passed through Trotter's accounts of which £15 million was transferred to his accounts at Coutts' bank. Despite the fact that he thought he was justified in using naval balances for his personal profit, he had qualms. In cross-examination by Mr Plumer, defending counsel for Lord Melville in the impeachment, Trotter said that he "might have looked upon it as a still greater act of delinquency" upon his part if he had had to make his profits from money taken directly from the Bank of England. It would appear to have been the practice to hold navy balances rather than to deposit them regularly at the Bank of England, and in that respect Adam Jellicoe also held balances for his own profit, and these became the source of loans to Cort.

Alexander Trotter must have become familiar with—or at least suspected—Jellicoe's transactions in 1788 and was probably the instigator of the action which Lord Melville reported to the Commissioners of Naval Enquiry as follows:[5]

10 Bankruptcy

10 July 1788, from the Treasurer of
the Navy to the Deputy Paymaster

Sir,
 I observe by the monthly accounts that the balance in your hands has remained considerably greater than it used to be formerly.

 It does not occur to me that there is any circumstance in the order of the Business of the Office to render the change necessary, I would therefore wish an explanation of it, and if it is not necessary, I must suggest to you the necessity of reducing it.

I am, your obedient servant,
Henry Dundas

Adam Jellicoe replied[6] on the same date:

Sir,
I have the honour...
In compliance with which I beg leave to say, I shall take the earliest opportunity of paying in all my Balance, for which there appears to be no immediate demand; but in case you think a security necessary for the responsibility of the Situation which I have the honour to hold under you, I beg leave to offer the inclosed, amounting to a larger sum than I can at any time hope to have in my hands unemployed
I have the honour to be with most perfect respect,
& most unfeigned gratitude, Sir,

Your most obedient servant,
Adam Jellicoe

Navy Pay Office, 10th July 1788

The 'inclosed' was the offer of the assignment of Cort's patents as a security. Lord Melville also disclosed to the Commissioners of Naval Enquiry in 1804[7] that, after the correspondence reported above:

This led me to have a particular conversation with him, when he confessed to me, that for a considerable number of years he had been embarrassed from some advance or engagements he had come under with a Mr. Cort; that from the high opinion I had of him ... I had no reason to doubt the fidelity of his representation; and it was my belief that if he had lived to superintend the business himself, and to bring the profits of the patent Mr. Cort had obtained to their proper bearing, that the loss might have been repaired.

On 11th August 1789 Parliament was prorogued and we presume Dundas had returned to Scotland. Trotter approached a Baron of the Exchequer for a writ of immediate Extent against his deputy, Adam Jellicoe. The Commissioners of Naval Enquiry recorded the event: "The solicitor of the

Admiralty being out of town, Messrs. Chamberlayne and White, on the application of Mr. Trotter, undertook to conduct the process on behalf of the treasurer against Mr. Jellicoe. The extents were issued on 29th, and Mr. J. appears to have died on the following day".

On his death, among Jellicoe's papers in his iron chest was a statement which he had written on 11th November 1782.[8] After stating that it had always been his practice to keep Navy bills in his iron chest equal to the value of public funds entrusted to him so that in case of death his balance might be paid immediately, he went on:

> I have always had much more than my balance by me, till my engagement about two years ago with Mr. Cort, which by degrees has so reduced me, and employed so much more of my money than I expected, that I have been obliged to turn more of my navy bills into cash, and, at the same time, to my great concern, am very deficient in my balance. This gives me great uneasiness, nor shall I live or die in peace till the whole is restored.
>
> But should my death be *sudden*, and before I can make up the public balance, I hope I shall not be reflected upon, as I have always meaned honestly and justly; and I must earnestly entreat the Treasurer, Paymaster, or whoever is most concerned, not to be severe in requiring the balance to be immediately restored, lest my character should be traduced, and my son Samuel (who I hope is in a way to make a comfortable provision for his brothers and sisters) and my family be ruined.
>
> I have considerably more than £20,000 engaged in the business with Mr. Cort and my son, and have paid almost all my engagements and acceptances. I expect the returns in bills of ironmongery wares will now come in very fast.
>
> Mr Weston . . . in case of my sudden death, will assist in raising money for me, as I doubt not Messrs. Bennett and Cure, Fenchurch Street, to save me from any imputation, and my family from being hurt; and they may be sure of being repaid by remittances from Mr. Cort and my son, so I hope my balance will shortly be made up; and as I have ever meaned to deal justly and honestly with the public and every private person, I again entreat that some favour may be shown to the family of one who served faithfully for above forty years, and it will not be insisted upon that the balance be immediately paid up.

On Monday 31st August the Commissioners and jury at the office of the Sheriff of Middlesex in Took's Court found[9] that Henry Cort and Samuel Jellicoe were indebted to his Majesty for the sum of £429 2s 4¾d "received by John Whiteway, Collector of H.M. Customs, Portsmouth [for] condemned goods sold at Portsmouth . . . and for which Cort and Jellicoe gave to the Collector a Bill of Exchange drawn on Adam Jellicoe . . . which had been returned".

10 Bankruptcy

On Tuesday 1st September the Commissioners and jury at Took's Court stated: "Whereas we are informed that Henry Cort is indebted to us in divers sums of money for as much of our money paid by Alexander Trotter of the Navy Pay Office Esquire Paymaster of our Navy unto Adam Jellicoe *late* of Highbury Place in the Co. of Middlesex Esquire . . . the said Adam Jellicoe at different times lent and advanced unto the said Henry Cort but no part of this money hath been paid by him to our use"—and they were to enquire and report before the Barons of Exchequer on 6th November. On the second skin the finding of the Commission was that "Henry Cort . . . is just and truly indebted to his Majesty in the sum of £27,500". Further inquisitions concerned returned bills of exchange drawn on Adam Jellicoe by Cort and (S.) Jellicoe. An Inquisition at Gosport on 31st October 1789 before the Sheriff of Hampshire and 12 jurors indicated the inventory of stock of Cort and Jellicoe ". . . all of which goods, chattels and debts were seized". There were six schedules, showing the value of stock:

Schedule 1 Gosport	£7,207	13 1
2 Fontley	3,984	13 1¼
3 Little Fontley Farm	518	0 0
4 Fareham	382	8 5
5 Sloop and 2 small boats	157	0 0
6 Debts owing	2,269	9 11
	£14,519	4 6¼

The Fareham property was probably a stocking ground on Portsmouth Harbour, for the stock consisted mostly of coal, scrap iron, firebrick and wood. The Gosport warehouse contained nearly £4,000 of manufactured goods: anchors, "nails for Dockyard", mooring and buoying chains, hooks, merchant nails, staples, horseshoe moulds, locks, files, twists, screws and augurs of wrought iron, some steeled weights and furnace bars of cast iron. There was also about £2,250 worth of unused iron: Russian bar, English square and round bars, plates and bolts. The remainder of the stock consisted of equipment in the various 'shops' and storehouse. The shops were for smiths with hammers, anvils, hearth plates and tuyeres. Bellows to the value of £171 15s 6d were collected under one heading, as were cranes (£30 5s 0d).

Since Fontley mill was not vacated by the last Gringoe until 1772, when Henry Cort took the lease, Gosport 'warehouse' had apparently been also the smithy of Messrs Attwick and Morgan when they held contracts to supply ironmongery to Portsmouth Dockyard.

61

Schedule 2, for Fontley (Titchfield), is of particular interest for it shows that the mill comprised:

	£	s	d
(i) The Mill with rollers, cutters, etc.	338	19	11
(ii) The Forge with hammers	171	12	1
(iii) The Smiths' Shop with hammers and anvils	77	17	4
(iv) The Finery with tools and boshes	18	15	9
(v) The Foundry	68	1	7½
	£675	6	8½

All except the foundry had cast-iron plates on the floors. There was no considerable stock of manufactured goods at Fontley but nearly £1,250 worth of new iron (wrought, including Russian), nearly £1,500 value of slabs and blooms (made from Welsh and Scottish pig, ballast, scull and burnt iron), some old wrought iron, some Carron and Ketley pig iron and some scrap cast iron. This schedule clearly shows that Fontley was the place where Henry Cort established the processes of both his patents. The mill had a considerable collection of rolls and slitting cutters (and steel for re-steeling the cutting disks) for slitting the rolled plates for faggoting, an essential stage in his processes.

One can see that a considerable amount of property was seized. On 27th November 1789 the Sheriff of Hampshire and a jury held an inquisition at the "inn called sign of the George in New Alresford" under the Writ of Extent issued by the Court of Exchequer on 29th August against the estate and effects of Adam Jellicoe, which resulted in his being seized of the freehold warehouse at Gosport let to Cort and Samuel Jellicoe, a freehold tenement and dwelling house on Portsmouth Common, a tenement, farm, etc. at Shedfield, and various debts due to him. Further properties in Islington were seized by the Sheriff of Middlesex, along with "certain Letters Patent (dated 17th January, 1783) to Henry Cort of Fontley ironworks which, by an Indenture of Assignement dated 29th August 1783 made between Henry Cort and Adam Jellicoe, were assigned to the latter for all the residue of the term of 14 years 'which now is of the value of £100' ".

It must have been clear to Henry Cort that he could not pay the debt of £27,500 he was alleged to owe to the King and that the only way to avoid imprisonment was to apply for a Commission in Bankruptcy. The assets of Cort and Jellicoe, however, exceeded the debts claimed against them.

10 Bankruptcy

Of the £36,500 alleged to be the total debt of Adam Jellicoe to the King, £27,500 was allocated as owing by Henry Cort, leaving £9,000 said to be owing by Messrs Cort and Jellicoe, the latter figure being amended downwards to £7,573 10s 7¾d a few weeks later which was, in effect, the sum deposited by Samuel Jellicoe with the Under Sheriff of Hampshire, and which allowed various goods and chattels to be returned to him.

We do not know whence Samuel Jellicoe obtained the money to pay off the debts of Cort and Jellicoe but he would not have had much difficulty in doing so because the assets exceeded the debts. Henry Cort must have—through his assigns—assented, though as an undischarged bankrupt he could no longer trade. Samuel Jellicoe therefore "was put solely into possession of the Trade and Effects"[10] of Messrs Cort and Jellicoe and held them until his death. Apart from repaying the loan he would have to pay £375 p.a. in servicing the loan as well as £200 p.a. which he undertook to pay on account of a sum of £4,000 advanced by his father to him personally—so that he had to face difficult financial problems for some time.

In the *London Gazette* of March 1790[11] appeared a notice that the acting Commissioners in Bankruptcy had certified to the Lord Chancellor that Henry Cort had conformed in all things required, that his Certificate in Bankruptcy would be allowed and confirmed unless cause be shown to the contrary before 3rd April 1790. Henry Cort remained an undischarged bankrupt until his death.

There is only one further record in the Weale MS[12] for this period:

> In May 1790 when the Meeting took place to consider what steps were necessary to be taken respecting the Patents it was determined to lay by quiet to give people an opportunity to erect their works if they should be inclined to follow H. Cort's process and in ye mean time to procure privately information of what the ironmasters were doing. Mr. Cort on 17th May 1790 wrote to Mr Trotter offering his personal services to procure such necessary information to render the patents productive but not receiving any Answer Mr. Cort of course c^d not proceed to procure such information and ye only step he c^d take under his circumstances were to procure information from a Master Roller (Jn Swaine) whom Mr. C had planted at Coalbrook Dale in ye works of Mr. Reynolds for the purpose of setting a going ye Rolling of Bars, w^h has had the desired effect inasmuch y^t some works he informs Mr. C. have followed the example laid down by ye said Master Roller—the several letters from him will show that his informations are *hearsay* as from the nature of his employ c^d not go off ye spot and as by the preceding paragraphs w^h Mr. C. is inclined to give credit to from his

hearsay information yt 30,000 tons were made in 1790 and Mr. Thompson 20,000 tons more will be made within 12 months.

It appears necessary to know these large quantities are making & about to be made & by whom & ye exact methods used—so as to judge whether or not they are infringements on Mr. Cort's Patents. Mr. C. may gain this information previous to any alarm.

July 1791   H.C.

In July 1791, when the last note was minuted, Henry Cort was in lodgings at Devonshire Street, Queens Square, London, when and where a letter was addressed to him by the Secretary of the Navy Board, the letter being a typical example of official evasion:[13]

> Sir, The Commissioners of the Navy have received your letter of the 4th instant, and I have it in charge from them to acquaint you that your invention appears of that utility as to induce them to give encouragement to the manufacture of British iron, performed according to the methods that have been practised by you.
>
> I am, Sir, Your most humble servant,
> Jno. Morrison, Secretary

There is a tradition in Gosport that at this time Cort's family lived in a state of penury and hardship which, unfortunately, cannot be given documentary proof.

The next letter on Cort's behalf was written in 1794, addressed to William Pitt, Prime Minister, and signed by ten members of Parliament and five others:

> Sir, We take the liberty respectfully to state to you the unfortunate case of Mr. Henry Cort, late of Gosport, at which place he had, some time ago, iron contracts under Government, and among the rest a contract for making malleable iron with raw pit-coal only, and manufacturing the same by means of grooved rolls by a process of his own invention; we are sorry to add that, through the very great expense necessarily attending the prosecution of these important instruments, this gentleman failed, when on the eve of reaping the harvest of his patents, which were taken possession of under extents from the Crown.
>
> We have therefore been induced, not only from compassion, but from the good opinion we entertain of him, and from the great national benefits which now actually result from these his discoveries and improvements, to join with many others in a subscription to afford some temporary relief to him and his destitute family, consisting of his wife and twelve children, (nine of whom are wholly unprovided for, only one able to maintain himself). This, however, cannot possibly produce any other effect than merely to supply his present urgent necessities, without rescuing him from

10 Bankruptcy

that state of poverty and ruin to which we are very sorry to find his meritorious exertions have reduced him; a circumstance which will appear the more unfortunate, when it is considered that the same pursuits have been attended with the greatest success and profit to others who, availing themselves of his experience, have enjoyed all the advantages without encountering the labour and difficulties ever inseparable from the first practice of any new process. [They then recommended Cort for some Government appointment in the Dockyards, Customs or Excise] . . . in any of which departments we will confidently engage that Mr. Cort's abilities and conduct will be such as by no means to discredit our recommendation, or any other countenance which you, Sir, may have the goodness to show him at our instance.

We have the honour to be, &c, &c.

The result was not an appointment to a government post but the granting of a pension of £200 p.a. (although deductions of £50 were made for 'fees').Thus, five years after his bankruptcy, Cort and his family were rescued from their poverty by a pension and the subscriptions of his well-wishers.

Henry Cort remained in lodgings in Devonshire Street, Queens Square, until his death in 1800. He was buried in Hampstead churchyard where his tombstone records:

Sacred
to the Memory of
MR. HENRY CORT
of Devonshire Street St. George the Martyr
Queen Square, London who
departed this life 23rd May 1800
in the 60th year of his age.
He passed away a broken hearted man.

When Henry Cort died his pension ceased. On 6th November 1801 his widow sent a memorial to the Lords of the Treasury for assistance and Sir Andrew Hammond, Comptroller of the Navy, was asked to report, which he did on 24th November:[14]

. . . that it was owing to the persevering industry of the Patentee (in which he expended a large sum of money) that this useful discovery was brought to light and which has since proved to be a great national benefit . . . I have always considered the Extent itself as a severe incumbrance on Mr. Cort as it does not appear that he knew the money he was borrowing from Mr. Jellicoe was the property of the Crown . . . The extreme indigence of the widow and her family is well known and my opinion is that they are real objects of national relief.

The Treasurer of the Navy was authorised to pay Mrs Elizabeth Cort £125 p.a. (with deductions to yield £100 net). When she died in 1816 a pension of £25 6s 0d (reduced to £19 by fees) was granted to each of Cort's two unmarried daughters.[15]

The outcome of Richard Cort's appeal in 1855 was that on 20th June he was granted a pension of £50 p.a. chargeable to the Civil List. In 1856 the two unmarried daughters (like Richard, over 70 years of age) had their pensions increased to £30 and in 1859 to £50 p.a.

CHAPTER ELEVEN

The Cort Parliamentary Enquiry of 1812

ONE OF the more perplexing problems surrounding the Jellicoe affair is the valuation of Cort's patents at a mere £100 and the subsequent failure of the Navy to energetically collect royalties from users of Cort's processes to defray the outstanding debt. Continued official indifference to the merits of Cort's processes provides the subject of this chapter.

It is pertinent to note that although a large number of blast furnaces were built in Monmouth during the Napoleonic wars, the use of puddling furnaces and rolling mills was not a feature of practice there until these wars were nearly ended. The early castings trade of the iron industry in the Black Country was not displaced as its main feature until the nineteenth century had begun and it was only in Glamorgan—indeed, in Merthyr Tydfil—that Cort's processes were extensively developed during the last decade of the eighteenth century.

If we take the total production of pig iron in Merthyr Tydfil in 1796 as representing the mean annual production for the eight years before Cort's patents could have been operated before they expired in 1797–8, allow two tons of pig iron per ton of bar iron and 5s 0d per ton of product, the total royalty payable would have been only £16,000. Even if we double this to allow for possible royalties elsewhere, the total does not exceed the prodigious expenditure Cort incurred in developing his processes. Even if Cort's patents had been extended to a period of 31 years—as was done for James Watt's basic patent—the Monmouth development would have been outside this period and it would probably only have served to restrict the development in the Black Country. In fact Henry Cort, or rather his family (for Henry died soon after the period of his patents expired), could only have been compensated for the benefits his processes conferred on his

67

country by a government award. It is with the allegations that these did not deserve such an award, as the Select Committee of the House of Commons concluded in 1812, that we are concerned in this chapter.

The petition on behalf of the family of Henry Cort, consisting of a widow and nine other children, was offered to be presented to the House on 24th January 1812 and three days later was referred to a Committee which had power to send for persons, papers and records.

One of the key witnesses on 14th February 1812 was Samuel Homfray who, it will be recalled, was uncivil to Henry Cort throughout. Homfray, in answer to questions, contended that a process similar to puddling, called 'buzzing', had existed at Coalbrookdale and Yerton[1] before Cort's patents and that the improved quality of British iron at that time was due to the making of finer's metal before puddling. Questions relating to 'fluted rollers' elicited the reply that they had been previously introduced by Mr Butler of Rochester works near Newport in 1782 and prior to Cort's patent of 1784. On 17th February he was examined again and maintained that the plan detailed in Mr Cort's petition was not followed by any manufactory and that the production of finer's metal was not the same process as puddling, though he admitted that the making of finer's metal required a coal-fired furnace similar to that required for puddling. He added that he had seen a furnace charged with blooms from the coke fineries and rolled into bars fit for merchant service, previous to the years 1783 or 1784, but rolled into plates and not bars.[2]

This evidence was very damaging to Coningsby Cort's case, and needs to be examined.

We can take, in turn, the assertions of the prior use of 'puddling' at Coalbrookdale and Eardington.

The processes used by the Coalbrookdale group are known in more detail than those of other companies of the period. Joseph Reynolds (1768–1859), then the only surviving son of Richard Reynolds, in a letter to William Cort dated 20th February 1812[2] referred to Henry Cort's visit to Ketley in 1784 and said: "When they came to my father's house, a claim was set up, in opposition to Mr. Cort, that the process of puddling was not new, but that it had been done for years, at Colebrook Dale . . . Indeed, I apprehend, that a patent for making Iron in an air furnace was taken out by the Coalbrook Dale Co., and that this expires about, if not before, the date of Mr. Cort's patent for Pudling".

It is clear from this and other references that the Coalbrookdale group rested their objection to paying royalties on the patent of Thomas

11 Parliamentary Enquiry 1812

Cranage of 1766 *because he used an air furnace*—a reason which, of course, is inadequate. Samuel Homfray's reference to 'buzzing' is misleading, and probably referred to minor operations of making scraps into 'Buz Balls' and then into half-blooms, a process related to the old stamping and potting method; 'buzzing' is clearly not puddling as Homfray alleged. James Cockshutt, in a letter to William Cort on 20th February 1812, stated that he had been told about a similar method to puddling being tried at Coalbrookdale, but a visit to that works sometime after his Fontley visit provided no evidence of any such process being practised there, though he was on the most intimate terms with Mr W. Reynolds.

At Eardington, John Wheeler's forge was built after the first patent of Wright and Jesson of 1773 for a process of fining which was being used by them at their West Bromwich forge, and it is this process which was used at Eardington. Except for the certainty that coke was used in the preliminary fining, the process was very similar to that of the Charles Wood process.

The Eardington forge was, therefore, one of the '60 melting refineries for working with coke and stamping' which Mushet recorded as being in use in 1788.

Samuel Homfray had alleged that preliminary fining, or the production of 'finer's metal', was the essential basis for the success of the process carried out in South Wales, and one which Cort had not mentioned in either of his patents. Finer's metal was pig or cast iron which had been subjected to a preliminary treatment (melting in an oxidising atmosphere and sudden cooling) by fusion with coke in a 'refinery', tapping on to a plate and quenching. This was (we now know) a 'desiliconising' process but was then described by the obvious visual change from a grey to a white iron. This preliminary 'refining' was well known to Henry Cort who, in his notes dated January 1789 gave the costs of using both grey iron and refined iron, both of which were obviously being tried out at Cyfarthfa at a time when his patents were acknowledged. Although costs were higher for refined iron, the operation was subsequently more rapid, and this process in later years was to be closely associated with the puddling furnace; indeed "the pernicious cinder being first separated by a previous operation . . . an improved kind of Iron is obtained and also with greater despatch—so that, it appears to me, without the use of the Pudling furnace the discovery (if it may be deemed a discovery) of making run out [i.e. finer's] metal *would be of little or no advantage*". Thus wrote James Cockshutt to William Cort on 23rd

April 1812[3] from Wortley ironworks. This is confirmed by a letter to Coningsby Cort by Alexander Raby of Llanelly, dated 20th June 1812:[4]

> I soon became a convert to his [H.C.'s] Mode of Puddling and Rolling Bar Iron both as to quantity and quality. Some improvement soon after took place by first refining the pigs or Cast Iron before they puddled it, but it was obliged to be puddled before it was made into Bars, nor could their refined metal be worked to Profit or with Expedition by any other Mode than by Puddling—I will venture to say that owing to that system and the use of grooved rolls, this Kingdom is indebted for its present proved State of the Iron Trade when instead of being obliged to bow down by the North of Europe for a Scanty supply of that useful Article at a very high Price, we are enabled to supply all our wants with an immense Annual Saving . . . and with whom did it originate but with your father. Envious Persons may want to prevent the Merit that is due to him.

Homfray's assertion that Mr Butler of Newport had previously rolled iron is easily explained by the fact that one of Cort's workers, travelling to Wales from Fontley to collect his family, had "shown Butler's Man how to roll Bolts till which time they had made no Attempts whatsoever towards such a process"—quoting a further letter from Alexander Raby to Cort.

Three independent sources confirmed that Butler's use of grooved rolls was *after* the date of Henry Cort's patents, one pin-pointing it as December 1786. One of the Committee, Ben Hall, offered an opinion favourable to the value of Henry Cort's patents and authoritative statements in support of Cort and refuting Homfray's allegations were received, but apparently arrived too late to be heard by the Committee.

A further damaging piece of evidence was that of William Crawshay, son of Richard, who held three-eighths of the shares in Cyfarthfa works and who stated that "if his family had pursued the plans of Mr. Cort, his family would have been ruined".

It is worth noting that Samuel Homfray had introduced puddling and rolling about March 1789 with plans obtained from Cyfartha forge, where the installation had been personally supervised by Henry Cort and his son William.

It is difficult to understand why Samuel Homfray, who was well known for a litigious disposition, should have been so evasive and deliberately misleading to the Committee. Despite a document being read to the Select Committee in support of Cort, a resolution of a Committee of Ironmasters which met at Gloucester on 29th March 1811 and which stated that "the iron trade was greatly indebted to the late Henry Cort for his exertions in introducing the puddling process to public attention, and

11 Parliamentary Enquiry 1812

for his invention of grooved rollers for the manufacture of bar iron"[5] the Committee was able to report to the House on 24th March 1812:[6]

> It appears to your Committee, that these alterations in such an important branch of Trade and Manufacture, have been effected by the industry and talents of numerous individuals . . . among these Mr. Cort appears to have possessed a considerable share of merit; but your committee have not been able to satisfy themselves that either of the two Inventions claimed by him . . . were so novel in their principles, or in their application, as fairly to entitle the Petitioners to a Parliamentary reward: moreover . . . no good malleable iron can . . . with certainty be made . . . by the method claimed as the invention of Mr. *Cort*, unless the Cast Iron has previously been made into what is called Finery or Finers Metal, by a process since found out . . .

The report ended with the hope that the House would vote the Petitioner a sum to cover his expenses; in the event Coningsby Cort was left to defray his own expenses.

It is instructive to note the result of a similar petition[7] relating to the spinning mule of Samuel Crompton. Crompton's petition, presented to the House six weeks after that of Coningsby Cort, was to result in a grant of £5,000 awarded a week before it discussed Coningsby Cort's petition.

It may be that a misleading list of improvements before the Committee, Homfray's tendentious statements, and a desire of the powerful naval members of the Select Committee to avoid any resurrection of the naval scandals with which Henry Cort was unwittingly connected (in the Whig attacks prompted by the Tenth Report of Naval Enquiry and the Impeachment of Lord Melville), coupled with the seemingly weak personality of the chairman, Davies Giddy, combined to ensure that the cause was lost.

Cort may not have achieved any official recognition, but eventually his processes were widely adopted by the iron industry throughout the country.

CHAPTER TWELVE

Iron production in Wales and the Black Country

IN 1788[1] there were 105 charcoal forges in Great Britain with 208 fineries, or approximately two fineries per forge (or hammer). If we assume working for 30 weeks in a year, the average British finery made 2⅔ tons of *wrought iron* per working week; the Welsh fineries were smaller than the average, producing 2 tons per week and the Stour fineries were outstanding, producing an average of 4⅓ tons per week. The average annual production per forge was 156 tons, a figure used elsewhere to compare with a Cort forge. Each of Henry Cort's puddling furnaces (as we may now call them to avoid confusion) had a weekly production of wrought iron not less than an average British finery hearth and could operate for 50 weeks instead of 30 in a year, to produce nearly twice the yield of an old finery. In a charcoal forge the chief limitation had been the capacity of the hammer to forge to bar; with a rolling mill a ten-fold output was easily possible. With coal as the fuel in the puddling furnace, difficulty in obtaining fuel no longer imposed a limit on the number of furnaces; and with steam-engine operation more than one rolling mill could be erected on a single site.

Richard Crawshay's new rolling mill of September 1789 had eight puddling furnaces attached to it, each capable of producing 3 tons of wrought iron per week (1250 tons p.a.). Since the design at Cyfarthfa was entirely that of Henry Cort, this was his greatest achievement and such a unit became the new single forge of the revolution in iron.

Even for the old-style double forge making 500 tons p.a. a single blast furnace making 16 tons of pig iron per week (800 tons p.a.) would have been more than adequate. The new-style rolling mill and puddling furnaces demanded a production of 1600 tons p.a. of *pig iron* (32 tons per week), the production of two of the older furnaces of South Wales—for in 1785 the total production of pig iron there was 60 tons per week or 15 tons

12 Production in Wales and Black Country

per furnace at Hirwaun, Dowlais, Plymouth and Cyfarthfa. Thus one can see that Cort's processes allowed a substantial increase in scale, especially now that fuel was no longer a locational limitation.

The new large furnace at Cyfarthfa, with its two tuyeres, was part of a large capital expenditure costing £50,000 which had included enlargement of the forge and the building of a new forge on the site of the old cannon foundry.[3] The production of 7200 tons of pig iron by Cyfarthfa's three furnaces in 1796 may be assumed to have been attained in May 1793 when the cost and the building of the new furnaces was reported. On the basis of figures given by Henry Cort this would produce 3740 tons of merchant quality iron or 75 tons per week, two to three times the production of the first mill in March 1789. When Malkin visited the South Wales ironworks he said:[4]

> Mr Crawshay's Iron works, of Cyfarthfa, are by far the largest in this kingdom; probably, indeed, the largest in Europe; and in that case, so far as we know, the largest in the world. He employs constantly fifteen hundred men. He makes, upon an average, between sixty and seventy tons of bar-iron every week, and has lately erected two new additional furnaces, which will soon begin to work, when he will be able to make, one week with another, 100 tons of bar iron. Mr. Homfray makes weekly, on a moderate average fifty tons of bar-iron and upwards, and is now extending Penydarren and its buildings, which will soon be completed; he will make at least eighty tons per week . . . The number of smelting furnaces at Merthyr Tydfil is about sixteen. Six of these belong to Cyfarthfa works, the rest to the other gentlemen. Around each of these furnaces are erected forges and rolling mills, for converting pig into plates and bar iron.

The next series of records for the early years of the nineteenth century are ambiguous; there is a single early canal account[5] which showed that, in one year from 1st October 1805, Cyfarthfa works sent 9906 tons of 'iron' down to Cardiff but the type of iron is not distinguished. The official estimates of pig production for 1806 are for counties; the production at Cyfarthfa was said to be 10460 tons and at Penydarren 7803 tons.[6] Six furnaces producing 10460 tons of pig iron in 1806 do not compare favourably with three furnaces producing 7200 tons in 1796, and presumably only four were in blast in 1806. The 1806 list is the first which gives the total number of blast furnaces and the number in blast (one-third of those in South Wales being out of blast) which supports the assumption made.

There is no ambiguity about the figures given for 1813 by Wood,[7] who recorded that the amount of bar iron produced per annum was:

Cyfarthfa 11000 tons; Pennydarren 7000; Dowlais 5000; Plymouth 4000 tons—all sent down the canal.

Richard Crawshay died in 1810. During his lifetime, therefore, and in fact in a period of under 25 years, the coke blast furnaces at Merthyr Tydfil had increased their production to nearly fifteen-fold that of 1788. In the 20 years to 1810 the production of pig iron at Merthyr Tydfil was mostly converted into bar iron but in the other iron-producing areas cast iron was still the primary product in wartime. Although exact figures are not available for the production of bar iron, it is probable that in this period in the parish of Merthyr Tydfil it was equal to that in any English county. By this standard this was an amazing transformation and Merthyr Tydfil became the Mecca for travellers from the whole of Europe.

The prime reason for this transformation was the adoption of Henry Cort's rolling mill which could then (with steam-engine operation) roll to bars 100 tons of blooms per week—a rate twenty-five times that of a forge hammer. For this production Henry Cort's finery furnaces could be built in any required number, each using the cheap fuel, coal in its raw state, to which the formidable problems of purchasing and carriage for charcoal did not apply.

In Monmouth records are available from 1802 to 1840[8] and show that castings only formed about 2 per cent of production. The quantity of refined iron (finer's metal), puddled bars and blooms, and finally bars, rods and hoops made by Cort's processes gradually increased until, from 1813 to 1840, they formed from two-thirds to three-quarters of the products sent down the canal. By 1806 more pig iron was made in Monmouth than in Glamorgan, and this disparity continued to the end of the records quoted. With the development of fineries from about 1810 it appears that from 1820 to 1840 at least as much wrought iron was then made in Monmouth as in Glamorgan, though the defects of the classification of the goods on the Glamorgan canal prevent a more specific statement. In both counties the expansion of the iron industry was based on the application of Cort's process of fining, piling and rolling.

There was one other region in Great Britain where expansion of the iron industry followed the same explosive course as in Glamorgan and Monmouth, namely South Staffordshire—the Black Country, as it came to be termed. Pig-iron production in South Staffordshire exceeded that of either Glamorgan or Monmouth in the three years 1806, 1823 and 1830.[9] There were, in effect, two areas of development, each bounded by the outcrop of the Thick seam of coal: one between Darlaston, Bilston and

12 Production in Wales and Black Country

Tipton and the other in the south between Dudley and Stourbridge. When Cort's patents had, in effect, been cancelled and James Watt's basic engine patent expired (1800), the building of coke blast furnaces, finery furnaces and rolling mills became very rapid and 'Staffordshire', with its central position and superb canal system, became the premier iron-making county of Great Britain, at least in terms of overall production.

By contrast with the rapid development in Glamorgan, Monmouth and the Black Country, Shropshire, Scotland and the combined areas of Yorkshire, which had all developed their iron-making in the foundry trade, did not experience such an upward surge of production in war-time and during the post-Napoleonic war period. An exceptional ironworks in South Yorkshire was that of Low Moor (the foundation of the first coke blast furnace was laid in May 1790). Soon after the 40 ft blast furnace went into blast, the Napoleonic wars started and the ironworks were concerned in casting shot and shell, guns and mortars. In 1801 the first forge, the traditional charcoal finery, chafery and hammer, was built and made small bars, nail rod and horseshoes. By 1803 the process of puddling with a refinery was introduced and so began what became one of the most-famed applications of Cort's original processes. The first puddler, John Davies, was given a piece rate, house rent, and expenses for moving his family, probably from Merthyr Tydfil. In 1805 a plate-mill was added and the making of a plate 6 ft by 3 ft by $3/8$ in. on one heat was recorded as a novel achievement, as indeed it was.[10] Low Moor iron was to acquire a world-wide reputation with railway engineers, and was well known for coupling-chains and carriage and wagon draw-gear. Even when techniques were changing from the 1830s, Low Moor continued to produce Best South Yorkshire Iron by Cort's method of dry puddling. The last cold-blast furnace there was blown out in 1935.

CHAPTER THIRTEEN

Conclusion

WE HAVE taken the story of Cort's processes to the end of the Napoleonic wars, at the cessation of which a slump in demand hit much of the iron industry. There were, of course, to be modifications to Cort's processes but these did not essentially alter the nature of the practices which he had introduced. Cort's puddling process relied on a sand hearth, and was subsequently known as 'dry' puddling. The significant modification of this practice was 'wet' puddling or 'pig-boiling', developed by Joseph Hall from about 1814, but not widely introduced until ca. 1830. Wet puddling consisted of the encouragement of decarburisation of the pig iron by adding magnetic, or black, oxide of iron (conveniently obtained as hammer-scale in forging hot iron) to the charge of pig iron. Black iron oxide is a strong oxidant, and in reacting as such was aided by a series of liquid phases, at first by the molten pig iron flowing over it, secondly by molten ferrous silicate and finally when the black oxide itself melted and flowed over the solidifying iron. The oxygen the black oxide gave up in oxidising the carbon in the iron was conveniently replaced from the oxygen in the combustion gases so that molten iron oxide could continue to act as an oxygen donor. During the removal of the carbon, carbon monoxide was evolved from the substantially liquid and entirely plastic mixture which rose like dough and the evolution of gas caused eruptions on the surface which seemed to boil—hence the process was known as pig-boiling, or wet puddling, as compared to Cort's dry puddling, in which there were only films of liquid medium.

By the time Hall introduced his process at Bloomfield ironworks in 1830 the use of metal iron bottoms for puddling furnaces was becoming common, coated eventually with roasted puddling cinder (ferrous silicate) which he called 'bulldog'. It would appear that in different parts of the country wet puddling and dry puddling were practised concurrently, the former having an apparent advantage of speed while the latter may have had the edge on quality, as at Low Moor. Wet puddling also gave a

13 Conclusion

greater yield and could use grey or unrefined iron. Dry puddling usually relied on refined or white iron; nevertheless, by preheating the succeeding charge at the base of the puddling chimney it was possible in a dry puddling operation to make ten heats of refined iron, using charges of three cwt, in a shift of ten hours (South Wales).

There were also refinements in rolling. The puddling or roughing rolls had triangular grooves to give a triangular cross-section. In the six diamond grooves the size decreased for successive passes. After the last pass in the roughing rolls the product was transferred to the left-hand rectangular groove, then the second and so on until a mill bar of the required size was obtained. The rolls rotated in opposite directions; the bar was caught with a pair of tongs by a 'catcher' on the opposite side who returned it over the top, the revolution of the roll carrying it back to the feeder, and so on. The product was sheared into short lengths which were piled (faggoted), heated in a furnace to a welding heat, slabbed and rerolled in another forge-train; if desired, the sequence was repeated. With a puddle ball of 70 lb, or even two welded together, the length of the final products, plates, squares or rounds, was never very great except for small rod. This cooled so rapidly that the 'three-high' mill was developed to give an active pass both backwards and forwards. Such a three-high mill was seen at Bilston, Staffordshire, by Thomas Butler in 1815 and appears to have been a Staffordshire development. A further refinement in that county was the guide-box through which the rod was passed finally to overcome the 'fins' on the bars from the squeezing operation, which in Cort's time had been "smoothed up under a light hammer".

These, then, are some of the major developments in iron manufacture which were to transform the British iron industry. In the second half of the eighteenth century England relied on Russia and Sweden for two-thirds of her consumption of bar (wrought) iron, the imports rising to nearly 50000 tons p.a. but starting to decrease in 1798 to the level required by the special needs of the raw material for the Sheffield crucible-steel industry.

By 1796 the British production of pig iron had doubled compared with 1788 to equal that of its greatest rival, France; by 1806 the production had again doubled in Great Britain but remained stationary in France. Since exports of wrought iron increased from under 1000 tons in 1796 to over 22000 tons in 1810, it may be concluded that Napoleon's attempted blockade of the Baltic was frustrated. This is in no small part due to Henry Cort, whose processes were increasingly adopted during the 1790s.

In Great Britain, mild steel making developed at the expense of wrought iron production. By 1885 the production of steel equalled that of puddled iron and subsequently greatly exceeded it. While puddling was to decline, the steel industry adopted the current rolling practices which are still the basis of the modern production of steel.

Postscript

THE FAILURE of the appeal of Coningsby Cort to the House of Commons in 1812 for some national recognition of the merit of his father's achievements has been recorded earlier. This had one beneficial result because Cort's sons, Coningsby and William, collected rebuttals of the misrepresentations made by William Crawshay and Samuel Homfray. Between 1855 and 1859 Richard Cort, the last surviving son of Henry Cort, made a number of appeals on behalf of one widowed and two unmarried sisters. In June 1855 he wrote a series of articles in the *Journal of the Society of Arts* containing some of the rebuttals of the evidence of 1812. A 'Cort Testimonial Fund' was formed in London. On 20th June 1855, on the recommendation of the Prime Minister Lord Palmerston, Richard Cort was granted a pension of £50 in the Civil List and his two unmarried sisters had their pensions increased to £19 p.a. and eventually to £30 p.a. and, in 1859, to £50 p.a.

The next memorial is dated about 50 years later and was a personal contribution by an American, Charles H. Morgan (1831–1911), who had earlier been the superintendent of a Bedson continuous rolling mill, which he modified by means of guides which twisted a bar between each pass, his modified mill being adopted world-wide.

Charles H. Morgan presented a bronze bas-relief which was erected in the porch of Hampstead church and unveiled on 9th March 1905 by Sir J. William Benn, the Chairman of the London County Council, in the presence of representatives of the Society of Arts, the Institution of Civil Engineers, the Institution of Mechanical Engineers, the Royal School of Mines and the Iron and Steel Institute. In 1966, the Institution of Civil Engineers restored the lettering on the tombstone of Henry Cort in Hampstead churchyard. In 1983 the Historical Metallurgy Society will erect a plaque at a suitable site to commemorate the bicentenary of Cort's patents, about which Charles Morgan had the following to say:

> Every . . . invention is subject to claims of priority advanced in behalf of unsuccessful predecessors . . . but the principles of both English and

American patent law make short work with such pretensions. The law was not instituted to reward prophetic genius or intuition. It rests on the proposition that a man who has discovered and successfully practised an improvement in the arts, and who might keep it as a trade secret, to die with him, shall be induced, by the grant of a monopoly of a limited term, to tell his secret completely to the public, so that, after the expiration of that term, any expert in the art concerned be not only entitled but enabled to practise it . . .

On the whole, it may fairly be said, that beyond all honest doubt and by substantially universal acclaim, Henry Cort was the first to perfect, put into successful operation, give to the world by sufficient description, and teach to other operators, his licensees, the puddling of iron and the rolling of puddled iron between grooved rolls.

Perhaps the last word should rest with Samuel Smiles:

"While the great ironmasters, by freely availing themselves of his inventions, have added estate to estate, the only estate secured by Henry Cort was the little domain of six feet by two in which he was interred in Hampstead churchyard".

References

Special reference books
The abbreviation used is the name of the author and the date of his publication, e.g. *Lloyd, J. (1906)*

LLOYD, J., The early history of the old South Wales ironworks, 1906
MUSHET, D., Papers on iron and steel, 1840
PERCY, J., Metallurgy: iron and steel, 1864
PLOT, R., The natural history of Staffordshire, 1686
RAISTRICK, A., Dynasty of ironfounders: the Darbys of Coalbrookdale, 1953
SCHUBERT, H. R., History of the British iron and steel industry, 1957
SCRIVENOR, H., A comprehensive history of the iron trade, 1841
STRAKER, E., Wealden iron, 1931

General reference books (and abbreviations used)
Encyclopaedia Britannica, 11th edition, 1911   *En. Brit. (1911)*
Dictionary of national biography   *DNB*
Victoria county histories   *VCH (County)*

State Papers
Journal of the House of Commons   *JHC*
Parliamentary Debates   *Parl. Deb.*
Tenth Report of the Commissioners of Naval Enquiry   *10th Rep.*
Impeachment of Lord Melville in Complete Collection of State Trials, XXIX (1804–6), edited by T. J. Howell, printed by T. C. Hansard, 1821, pp. 550–1482   *Trials*

Journals, Transactions
Journal of the Iron and Steel Institute   *JISI*
Transactions of the Newcomen Society   *TNS*
Law Quarterly Review   *LQR*

Transactions of the Institution of Mining Engineers   *TIME*

Special collections

British Museum Library	*BM*
Public Record Office	*PRO*
National Maritime Museum	*NMM*
Weale, J., jun., Account of the iron and steel trade 1779–1805, Science Museum (*see below, Chapter One, note 10*)	*Weale MS*
Wood, Charles, An account of the material transactions at Cyfarthfa in the parish of Merthyr Tydfil commencing April 1766 (*see below, Chapter One, note 2*)	*Wood MS*

Boulton & Watt Collection, Birmingham Reference Library
Boulton Papers, Assay Office, Birmingham

Chapter One: *The search for new finery processes*

1 Although in this chapter the word *refining* is sometimes quoted for this process, the correct term is *fining in a finery*; as will be seen, a 'coke-fired refinery' was later used as a preliminary to fining

2 Charles Wood MS: two manuscript books totalling about 400 pages—*An account of the material transactions at Cyfarthfa in the parish of Merthyr Tydfil commencing April 11, 1766* and covering the period to May 1767 when Cyfarthfa forge was nearly completed and the building of a blast furnace had begun. In the back of Book 2 there is an account of Charles Wood's work at Lowmill, Cumberland, chiefly for the period 1752–4. The MS belongs to Wing-Commander F. J. P. Wood of Moorabbin, Victoria, Australia, who has kindly made a microfilm copy available

3 Wood MS: an interim report on the Wood family genealogy, 15th September 1963; it includes a genealogical history compiled by a great-grandson of Charles Wood, with other family records

4 The use of coal, however, often meant in practice coke made *in situ* as in the age-old process of the smith, and this was certainly the case in the finery processes of 1788

5 Mr Cockshutt would be John I (?–1765) of Wortley forge, near Sheffield, whose son John II (?–1798) in 1771 had a patent for making wrought iron in a finery using pit coal; James, his brother (after being a manager of Pontypool forge and rolling mill for 10 years) was in 1784 a partner in Cyfarthfa forge built by Charles Wood in 1766

6 This apparent draining of a final cinder was due to the oxidation of silicon to silica which reacted with iron oxide to form a slag or iron silicate; silicon was then unknown

References

7   The ridge separating the firing hearth from the furnace space
8   Two furnaces heated per day would be required to give the stated production of 10 tons per week
9   de la Houlière, Marchant, Report to the French Government: *Smelting iron ore with coal and casting naval cannon in the year 1775*. English translation by W. H. Chaloner, *Edgar Allen News*, 1949, **27,** 213
10 Weale, J. jun., MS *Account of the iron and steel trade 1779–1805*, Science Museum Library, now rebound in one volume but catalogued as two volumes; unpaginated; the author has assumed that the original Vol. I extended to ca. 1790 and that Vol. II deals with the Cort petition of 1811–12: these supposed or original volumes have been paginated by the author
11 *Ibid.*, **I,** 63
12 Mushet, D., *Papers on iron and steel, practical and experimental, 1841,* 44
13 Patent Specifications from the Blue Books of the appropriate year, first printed 1854. The dates are those of final enrolment; the earlier dates of filing are often given in the literature
14 *Weale MS,* **I,** 24
15 Smiles, S., *Lives of the engineers: iron workers and toolmakers* (1st edn 1863), 1905, 88
16 Morton, G. R., *The Metallurgist*, 1963, **2,** 259–68
17 Campbell, R. H., The Carron Company, 1961, 55–6
18 *Ibid.*, 58
19 *Boulton & Watt Papers*, Birmingham Public Libraries
20 Mott, R. A., *TNS*, 1959–60, **XXXII,** 50–2
21 *Weale MS,* **I,** 17–18
22 Percy, J. (1864), 639
23 Schubert, H. R. (1957), 284–5

Chapter Two: *The background of Henry Cort: Navy Agent*

1   Giles-Puller Collection 1763–1831, County R.O. Hertford, A903
2   *Ibid.*, A904–6
3   Allegations before the Surrey Commissary Court, Greater London R.O.
4   Goss, C. W. F., *London Directories 1672–1855*, 1932
5   *Weale MS*, Science Museum Library, **II,** 35
6   *Mechanics Mag.*, 1859, **II,** 36
7   Giles-Puller Collection, A907
8   Hulme, E. W., *Notes & Queries*, 1952, **197,** 77–82
9   Dickinson, H. W., *TNS*, 1940–1, **XXI,** 31–47
10 National Maritime Museum Library, Greenwich, *POR/A*, various

Chapter Three: *The early history of Fareham Ironworks: Fontley and Gosport*

1   *En. Brit.* (*1911*), Anchor, Admiralty, Portsmouth
2   *En. Brit* (*1911*), Earls of Southampton, Dukes of Beaufort and Portland; Burke's *Peerage*, the same
3   Information from the Archivist, Hampshire Record Office, Winchester

83

4 *Receiver General Accounts*, 5M53/950, f6v, Hampshire R.O.
5 *Ibid.*, 5M53/767, 29
6 *Ibid.*, 5M53/768, 77–78
7 Court Books, 5M53/741, f36
8 *Receiver General Accounts*, 19M48/37
9 Hulme, E. W., *TNS*, 1928–9, **IX,** 12
10 *Court Leet Petition*, No. 79, Southampton R.O.
11 Information from the Archivist, Beaulieu Abbey
12 Fell, A., *Early iron industry of Furness and district*, Ulverston
13 Auty, B. G., *Trans. Hist. Soc. Lancs.& Cheshire*, 1957, **109,** 108
14 Editor's note
15 *Weale MS*, **I,** 76–9

Additional note: Fontley Mill appears to have the honour of being the first recorded tin-plate mill in the country (letter of John Tilte, 1623, Kidderminster Library, Knight MSS): "The Mill which batters the iron is the Earlles of Southamptons...". In 1628 a (tin) plate mill belonging to the Earl's neighbour at Wickham, just up-stream from Fontley, was leased for 21 years to Thomas Jupp, girdler, of Crooked Lane, London. 'Crooked-Lane ware' consisted of kitchen utensils and the like made from tinned iron plate. In 1661, Dud Dudley and William Chamberlaine took out a patent for tin-plating; the Chamberlaines were stewards to the Earls of Southampton in earlier times.

The Gringoes were Quakers. A Roger Gringo and his wife were buried in 1703 and 1671 respectively at the Friends Burial Ground at Swanmore, not far from Fareham. Another Roger, of the Ringwood Quarterly Meeting, was buried in 1693 at Ringwood, and would have been connected with the Sowley arrangements. John Gringo, mentioned in relation to Bursledon, died at the age of 84 in 1776; he may not have been a Quaker (Hants & Dorset Registers; Friends House, London). It would be interesting to know their connections with other Quaker iron enterprises. The Fontley site was excavated 1975–76 (P. Singer; forthcoming) [*Ed.*]

Chapter Four: *The development of puddling and rolling at Fareham*

1 *Boulton & Watt Collection*, Birmingham Reference Library
2 *Weale MS*, **I,** 76–9
3 Scrivenor, H., *A History of the Iron Trade*, 1841, 405–6 for prices of Swedish bar 1782–1812
4 Webster, T., *Mechanics Mag.*, 1859, **II,** 53. The date of the agreement with Adam Jellicoe was 8th January 1781, Jellicoe to be entitled to half the profits and paying half the costs of the contracts and stock-in-trade at a price to be settled by arbitration

Chapter Five: *The rolling of metals*

1 *Phil. Trans.*, 1712, **27,** 467–9, letter Edward Llwyd to Tancred Robinson, Usk, 15th June 1697

References

2 Gibbs, F. W., *Annals of Science*, 1956, **II**, 145–52
3 Smeaton's designs, Roy. Soc. Lib., II, f, 70v; Newcomen Soc., *Extra Publication* No. 5, 37
4 Johannsen, O., *Geschichte des Eisens*, 1951, **263–4,** 524
5 Translated by Akerman, R., in Swank, J. M., *History of the Manufacture of Iron in All Ages*, Philadelphia, 1884, 69–70; see also Rhodin, J. G. A., *TNS*, 1926–7, **VII,** 17–23, and Hall, J. W., The Making and Rolling of iron. *TNS*, 1927–8, **VIII,** 40–55
6 Gale, W. K. V., *TNS*, 1964–5, **XXXVII,** 35

Chapter Six: *Henry Cort's puddling process*

1 Patent Specifications from the Blue Books of the appropriate year, first printed 1854; the dates are those of final enrolment
2 'Plastic' would be more correct

Chapter Seven: *Testing Cort's iron in the Naval Dockyards, 1783–6*

1 Webster, T., *Mechanics Mag.*, **II,** 53 (re 1783)
2 *Ibid.*, 85 (re 1784)
3 *A brief state of the facts relating to the new method of making bar-iron with raw pit coal and grooved rollers discovered and brought to perfection by Mr Henry Cort of Gosport to which is added an appendix containing observations of Lord Sheffield for that subject and letters approving Mr Cort's method from David Hartley, Esq., Dr Black Professor of Chemistry at Edinburgh and others.* March 1787, 17 pp.; a copy marked 'Lord Sheffield from the author' with the original letter from Cort dated 10th November 1787 is in the *Weale MS*
4 Scrivenor, H. (1841), 150–1
5 *Weale MS*, **II,** 2
6 *Ibid.*, **I,** 80–1
7 *neutral* = wrought; *Sweeds* = Swedish
8 Ref. 3, General Remarks, 11–12
9 *Ibid.*, 1–10
10 Ref. 3, 13; Webster, T., *loc. cit.*, 101.
11 *Ibid.*, 16; Webster, T., *loc. cit.*, 101
12 'Sulphur' is incorrect; it was carbon, not yet so-called, which burned with a blue flame
13 *Weale MS*, **I**

Chapter Eight: *The adoption of Cort's processes (1) Coalbrookdale*

1 *Weale MS*, **I,** 17, 18
2 This is an overstatement for the period
3 *Weale MS*, **I,** 23
4 *Ibid.*, 24

Chapter Nine: *The adoption of Cort's processes: (2) developments at Cyfarthfa 1787–9*

1 *Weale MS*, **I**, 24
2 The Rotherhithe Company appears to have been the first works to adopt Cort's process and pay a royalty (29.5.1784). Cort occasionally visited the works, which appear to have operated his process successfully at a time when Cyfarthfa was having 'teething troubles' in installing Cort's processes, resulting in poor quality iron [*Ed.*]
3 *Weale MS*, **I**, 59
4 *Ibid.*, 60
5 'The former method' being 'potting and stamping'
6 This is at the rate of 750 tons p.a., more than double the production of the average of the large forges on the river Stour in 1788
7 At Portsmouth or Fareham [*Ed.*]
8 *Weale MS*, **I**, 62
9 *Ibid.*, 63
10 Editor's comment, after Hyde, C. K., *Technological Change and the British Iron Industry 1700–1870*, 1977, which examines comparative variable costs in Ch. 5
11 Near Derwenthaugh, where William Hawks installed a Cort rolling mill

Chapter Ten: *Bankruptcy*

1 10th Rep., App. 39, 425 (BM)
2 Trials, 902–3
3 10th Rep., App. 48, 441 (BM), Parl. Deb., **3**, 1200
4 Trials, 1201
5 10th Rep., App. 51, 452 (BM)
6 *Ibid.*, App. 52, 452 (BM)
7 *Ibid.*, Parl. Deb., **3**, 1189–90
8 *Ibid.*, App. 48, 441 (BM)
9 P.R.O., E.144/31
10 10th Rep. 985.q.13, App. 47 (BM)
11 *London Gazette*, Number 13182, Tuesday 9th March to Saturday 13th March 1790
12 *Weale MS*, **II**, 11–12
13 Webster, T., *Mechanics Mag.*, 1859, **II**, 388
14 Dickinson, H. W., *TNS*, 1940–1, **XXI**, 31–47
15 Cort, R., *J. Soc. Arts*, 1855, 634

Chapter Eleven: *The Cort Parliamentary Enquiry of 1812*

1 Eardington, Yerdington
2 *Weale MS*, **II**, 58–60
3 *Ibid.*, **II**, 26
4 *Ibid.*, **II**, 63

References

5 Cort, R., *J. Soc. Arts*, 1855, 608; Percy, J. (1864), 633
6 Parl. Deb., 1812 (118), **II**, 85–6
7 *Ibid.*

Chapter Twelve: *Iron production in Wales and the Black Country*

1 Mushet, D. (1840), 44
2 Coxe, W., *An Historical Tour in Monmouthshire*, 1801; Lloyd, J. (1906), 162
3 R. Crawshay to the Earl of Liverpool, 6th May 1793, quoted Addis, J. P.
4 Malkin, D. H., *The scenery, antiquities and biography of south Wales*, 1st edn 1804, 1807, **I**, 264, quoted Lloyd, J. (1906); Scrivenor, H. (1841), 122
5 *Ibid.*, 2nd edn, 1807, **I**, 204
6 Scrivenor, H. (1841), 97–8
7 Wood, J. G., *The principal rivers of Wales*, 1813; Lloyd, J. (1906)
8 Scrivenor, H. (1841), 127, 294
9 *Ibid.*, 97, 132–3; in 1806 "south Wales" included Monmouth, but records of the products carried on the Monmouth canal enable the total to be split
10 Mott, R. A., *TNS*, 1959–60, **XXXII**, 43–51

APPENDIX I

The wider significance of Cort's processes

THE IMPACT of Cort's processes on the wider economy after the 1790s has often been underestimated.

With the exception of two modifications in the design of the puddling furnace, the techniques developed by Cort remained unchanged between 1791 and the late 1830s. At the same time there was, during the period of war (1790–1815), an encouragement to the trade as a result of increased duties on imports, and the economies of scale resulting from concentration on the coalfields ensured falling costs. Forge profits were high during this period but declined in the 1820s and '30s as prices fell against an accompanying inability to increase productivity.

Finished bar iron was drawn into many uses—in agriculture for tools and equipment; in construction, as in the new mills of the period; for canals; for the great herds and flocks of the droving trade; in the fabrication of steam boilers for all purposes; for ships' plates; for domestic use; in the coaching trade—but, perhaps above all, in the construction of the railways, where the provision of heavy wrought iron rail during the first railway boom of the 1830s kept many a puddling forge in full production. While railways accounted for only 7% or so of total iron output, the increased scale of production appeared dramatic to contemporaries.

It is difficult, with few reliable estimates before the 1850s, to provide an accurate assessment of bar iron output. Hyde (*op. cit.* Ch. Nine), computing scattered data, provides the following:

Estimates of British bar iron output 1788–1815 (tons)

1788	32 000
1805	100 000
1810	130 000
1815	150 000

Cort's processes

These figures reflect the increasing adoption of puddling. By 1827, one can estimate an annual production of between 338 000 and 354 000 tons, an increase of some 130% since 1815. Since a considerable proportion of this output was now exported, where iron was previously imported, one can see how important were Cort's ideas to the economic development of the period.

APPENDIX II

Henry Cort's Patents 1783 and 1784

A.D. 1783 . . . . N° 1351.

Preparing, Welding, and Working Iron.

CORT'S SPECIFICATION.

TO ALL TO WHOM THESE PRESENTS SHALL COME, I, HENRY
CORT, of Funtley Iron Mills, in the Parish of Titchfield, in the County of
Southampton, Esquire, send greeting.

WHEREAS our Sovereign Lord King George the Third, by His Royal Letters
5 Patent under the Great Seal of Great Britain, bearing date at Westminster,
the Seventeenth day of January, in the twenty-third year of His reign, upon
the Petition of me, the said Henry Cort, whereby I had humbly represented to
His said Majesty that by great study, labour, and expence, in trying a variety
of experiments, and making many discoveries, I had invented and brought to
10 perfection a "PECULIAR METHOD AND PROCESS OF PREPARING, WELDING, AND
WORKING VARIOUS SORTS OF IRON, AND OF REDUCING THE SAME INTO USES BY MA-
CHINERY; A FURNACE, AND OTHER APPARATUS, ADAPTED AND APPLIED TO THE SAID PRO-
CESS;" and His said Majesty, being willing to give such encouragement as in the
said Letters Patent is mentioned, and having graciously pleased to condescend
15 to my request, made known, therefore, that His said Majesty, of His especial
grace, certain knowledge, and meer motion, for Himself, His heirs and
successors, did give and grant unto me, the said Henry Cort, my executors,
administrators, and assigns, His said Majesty's especial licence, full power, sole
priviledge and authority, touching my said Invention, as in the said Letters
20 Patent is in that behalf particularly mentioned and expressed; to have, hold,
exercise, and enjoy the said licence, powers, priviledges, and advantages
therein granted unto me, the said Henry Cort, my executors, admors, and
assigns, for and during and unto the full end and term of fourteen years from
the date of the said Letters Patent next and immediately ensuing, and fully to

A.D. 1783.—Nº 1351.

Cort's Method of Preparing, Welding, and Working Iron.

be compleat and ended according to the Statute in such case made and provided, to be had, used, and exercised, according to His said Majesty's gracious Intention, under His said Majesty's Royal Commands, within that part of His said Majesty's Kingdom of Great Britain called England, the Dominion of Wales, and Town of Berwick-upon-Tweed, conformable to His said Majesty's gracious pleasure, and to the provisoes, declaration, restrictions, grants, and conditions in the said Letters Patent expressed and contained, as in and by the said Letters Patent, so under the Great Seal of Great Britain, and dated as aforesaid, relation being thereunto had, may and doth more fully and at large appear.

NOW KNOW YE, that I, the said Henry Cort, in conformity to the said Letters Patent, and according to a proviso therein contained for that purpose, have described and ascertained, and by these Presents do describe and ascertain, the nature of my said Invention, and in what manner the same is to be performed (that is to say) :—

For large uses, such as shanks, arms, rings, and palms of anchors, mooring and bridle-chain links, and the like (in the doing of which I work with a forge hammer), various sorts of iron are prepared, welded, and worked, according to the said Invention, in the manner best suited to each particular use, as follows :—For instance, when a chain, link, anchor, ring, or other long piece of iron, of equal size at both ends, is to be wrought from one faggot, I cause barr iron of sufficient lengths to be closely faggotted together ; and the best method of doing this I have found by experiment to be by laying a flat bar, of the width and length required at the bottom, upon which four square barrs are to be placed in regular order, and upon the top and at each side other flat bars, perfectly inclosing the four square barrs, and thereby making the faggot square ; the size of all which bars are to be proportionate to each other, as well as to the use intended to be finished from them ; and the faggot so formed is to be bound together as close as possible by square collars driven on tight. But when the use intended is of considerable length, and tapering, so as to be smaller at one end than at the other, like the shank or arm of an anchor, to be wrought in its whole length from one faggot, I find that the same may be best and most conveniently formed by flat bars forged on purpose, gradually tapering so as to be thinner and narrower at one end than at the other, the proportions of such bars being calculated to correspond with the dimensions of the respective parts of the uses to be finished from them ; and in preparing such faggots these flat bars are laid over one another in the manner of bricks in building, so that the upper barrs everywhere cover the joints of those underneath ; one bar wider, than those which compose the body

Cort's Method of Preparing, Welding, and Working Iron.

of the faggot, being laid at the bottom, and another of the same width at the top, by which means the faggot, when taken from the furnace, is more easily rested on the anvil, and so placed as to receive the better impression from the first strokes of the hammer, whereby it becomes more perfectly welded. And
5 I observe the same method as nearly as possible in laying old hoops together on the building bench, as herein-after mentioned, and in preparing and faggoting all other sorts of iron for the welding furnace, as far as the shape of such iron and of the uses intended to be finished from the same will allow; several of the faggots so prepared, to the amount (for instance) of half a ton,
10 more or less, are put at once into a common air furnace, usually denominated a balling furnace, and therein brought to a welding heat at the same time, which in my furnace is compleated with great speed and dispatch; and this is a practice not hitherto used in any chafery or hollow fire, or other fire blown by blast, in welding large faggots of iron; and thus to take welding heats
15 in a balling furnace according to the method and process herein specified is part of my Invention. If the whole length of the faggots when prepared is too much to admit of being introduced entire into the furnace, the faggots may be first put in at one end, and, the other ends remaining out, the door is to be let down upon them, and closed at the bottom with sand; and afterwards,
20 the effects of the first heat being produced, the other ends are heated in like manner. But for the shanks of large anchors made entire in one length; and other faggotted iron of great length, for instance, of more than eight or ten feet long, then, in order to take the welding heat in the middle, a door is made in the back of the furnace, directly opposite to the one in front; and the ends first
25 welded being put through the furnace and out at that door, the same is let down and closed with sand in like manner as the front door, and the middle part only of the iron is confined to the heat of the furnace. But, for those purposes which require such a length, the mode I prefer is to weld the iron in two pieces, and afterwards to shut it together with scarfs of three or four feet
30 long, according to the plan herein-after more particularly described. But in the furnace used after the manner above mentioned, a welding heat may be taken much sooner than in a hollow fire, and that on lengths, for instance, of four feet or more at one time; and when the heat is perfect, the faggots or pieces heated and brought, either one by one, or two or more, to be
35 welded together, as the case may require, under a forge hammer of great weight, for example, eight hundred or nine hundred weight, more or less, the surface of which, and the anvil under it, is of sufficient dimensions, for example, of eighteen or twenty inches, more or less, in length, and about ten inches wide, except at the points, which are made narrow for the purpose of

Cort's Method of Preparing, Welding, and Working Iron.

drawing the iron out; and by a few strokes of this hammer, which acts with great velocity) striking on each side of the faggot (whilst the heat remains perfect), the whole length heated is welded round to the center into one solid mass of iron. For square uses, or those in which only the angles are beat down in finishing after the faggot is welded, the hammer and anvil just described with plane surfaces are the fittest to be applied; but for uses that are to be made round, such as anchor rings and the like, I work them at an anvil with a semicircular groove, and sometimes also by a hammer with a similar groove. When faggots are too heavy to admit of this operation being performed by hand, a crane is necessarily used; but two good forgemen can manage a faggot of three hundredweight in hand (by the assistance of others with barrs in lifting it from the furnace to and upon the anvil), and perform the work in general better than by the crane. And I find it to be the best way of making shanks for large anchors to lay and weld them in two pieces, one of the faggots being calculated for the square, and the other for the round part; and, after these are drawn out seperately, to shut them together under the said heavy hammer with a scarf of sufficient length, for instance, of three or four feet long, by a welding heat in the furnace. These shuts are much longer than those usually taken in a hollow fire or chafery, and, being perfectly sound, the anchors are altogether as strong and more neatly finished than those for which the faggots are at first formed of the entire length of the shank. I also remark, that by a welding heat from the furnace palms may be brought on the arms of anchors under the forge hammer with greater certainty and better effect than by the methods in practice before my Invention. If, instead of bar iron to be welded for the purposes as above stated, old hoops or other such like iron is to be used, the same are to be cut with the mill sheers or otherwise, or folded up into proper lengths for the intended uses, and then to be faggotted together by means of the common bundling bench, or by being driven into collars of the size and shape required, in both which cases I cause them to be laid as nearly as possible in the form herein-before specified with respect to barrs of iron, so as for the hoops of the upper layers to cover the joints of the under ones; and, a welding heat being taken on them, they are at once welded and drawn out to convenient dimensions under the large hammer. And for purposes that require two or more of these to be welded together, as shanks and arms, or bodies and palms of anchors, or mooring and bridle-chain links, and the like, they are first shingled sound apart, and then either returned into the same furnace, or (which I find better in works sufficiently large) put into another furnace of the same kind, and, being again heated to a welding heat, two of them are welded sound together,

A.D. 1783.—N° 1351.

Cort's Method of Preparing, Welding, and Working Iron.

and drawn to the proper dimensions and shape; and a third and fourth, and so on, may be afterwards shut on in like manner, after which the whole is neatly finished under the large hammer without the use of either chaffery or hollow fire. And I find that the larger uses thus finished by me are, in all
5 respects, possessed of the highest degree of perfection. I also have found that the fire in the balling furnace is better suited, from its regularity and penetrating quality, to give the iron a perfect welding heat throughout its whole mass, without burning in any part, than any fire blown by a blast; insomuch that if barr or other massy iron is bound into a faggot of five or six inches diameter,
10 by an old cask hoop the sixteenth part of an inch in thickness, the one will acquire a compleat welding heat, and the other not be burnt or wasted; and the consequence of this is, that whereas bar iron, faggoted and made into anchors, and other large uses, in the usual method practiced by anchorsmiths, is not welded to the center, but rather covered only with an incrustation of
15 welded iron when the anchors and such other uses made by them are compleatly finished, and consequently liable to decay and many accidents, the same large uses prepared by my method, whether from bars or from smaller iron, are equally solid through every part, and must necessarily be stronger, as well as less liable to injury, than iron not so perfectly compressed; and the whole
20 mass may, in one welding heat thus managed, be drawn out more or less in proportion to the solid contents of the length heated; on which account I always make my faggots shorter than the uses to be finished from them, and according to the usual course of my working I find that a square faggot of six feet six inches long, and five inches diameter, will draw out to about
25 eight feet in length, and a diameter of three inches and seven eighths, or thereabouts, in two heats, and the like. Old salt-pan plates and all other rolled plate iron cut with the mill shears or otherwise, and burnt together by itself or folded into a sort of coffins, and filled with scraps commonly called nut and bucket iron, which scraps may likewise be put into the furnace in piles, and
30 also lump and large iron may, for the like purposes, be welded by the same process, and worked under the forge hammer by means of porters welded to them, as practiced in drawing out blooms; but inasmuch as no hammer can press out all the earth, salt, and sulphur with which the iron may be overcharged if put into the furnace in a dirty, impure, or rusty state, at least not
35 so perfectly as may be done by the process herein-after mentioned, the same may be cleaned before being bundled for the furnace by a machine well known in the iron manufactory under the name of the scowering barrel, into which the larger iron may be put with scrap iron instead of small pebble stones, as used at wire mills, and the whole will come out perfectly cleansed,

A.D. 1783.—N° 1351.

Cort's Method of Preparing, Welding, and Working Iron.

and consequently, when wrought by the above process, will produce iron of a purer grain and more perfect contexture than without such cleansing; also by another process through the same kind of furnace, all the sorts of iron suited to the former operation being bundled or placed together as nearly as may be, with the care herein-before expressed, in such shapes as best to suit the uses for which they are intended, with one end or side thinner than the rest to give a free admission between the rollers of a common rolling and slitting mill, but not in greater lengths than can be heated entirely at once, or heated in like manner to a welding heat, and then passed through the said rollers; and by this simple process all the earthy particles are pressed out, and only such a proportion of the salt as is very much fixed, and of the sulphurious particles as rather invite the pure metalic grain into closer contact than promote its seperation, are left remaining, whereby the iron becomes at once free from dross, or what is usually called cinder, and is compressed into a fiberous and tough state. Accordingly, a bar of the worst ordinary iron, in which the grains of metal are large, and much seperated by the impurities with which it is impregnated, being passed through the simple opperation, becomes instantly of a good quality; and two or more of the same barrs, or of any pieces of impure iron heated in the same manner, and passed through the rollers together, become at once welded into one solid body, and meliorated into good tough iron, without being previously cleansed in the scowering barrel or otherwise. If the iron to be welded by this method is intended for plates of any kind, it is bundled or laid together, before being put into the furnace, in the shape and size best suited to the use for which such plates are intended, and passed repeatedly through the common rolling and slitting mill rollers with flat surfaces, till it acquires the requisite degree of thinness. If it be intended for flat or square bars, rods, hoops, or the like, as soon as welded and made sufficiently thin under the rollers, it may at the same heat be passed through the slitters, and from them again through the rollers, if needful, for proper uses. When the iron thus to be welded through the rollers is designed for barrs, half flats, or thimble iron, with feather edges, a groove of the requisite dimensions is made in the under roller, through which groove the iron is to be passed in the state of welding heat. In case thick bars, or squares, or round bolts are intended to be welded through the rollers, grooves of the shape and dimensions required for any of those uses are made in the under roller and collars on the upper roller to work exactly within such grooves, the surface of such collars being either plane for squares and flats, or concave for bolts and the like, as the case may require. Much of the success of these opperations depends on the iron being properly cut, faggoted, bundled, or placed together, if large, or

being put into coffins, as before described, suited to the intended purpose, if it consists of scraps, before being carried into the furnace; but this circumstance being attended to, and the coffins of craps, and even piles of the same, as well as faggots of suitable dimensions, being passed through any of the said
5 grooves, produce the intended use perfectly welded at the edges and throughout. And in mills of sufficient power, with rollers large enough to admit of grooves and collars for the purpose, very great-sized iron may be welded in this manner as well as under the hammer, the heat being the same in both cases, which process is my own Invention, and put in use by me only. I have
10 made faggots of inferior bar iron, five inches in diameter, and, passing the same through the groove in the rollers, have found them compleatly welded at the sides, without a crack, into one mass, perfectly sound to the centre, and improved in quality, in like manner as mooring-chain links, ships' knees, and other iron decayed or eaten by rust, being cut into proper lengths, duly heated and
15 passed through the rollers, will produce exceedingly good iron without any other process.

Another process relates to the management of scull and cast iron, the scull iron being first piled together in heaps upon old plate or bottoms, for example, of about three quarters of an hundredweight, more or less each, and shingled
20 and drawn down by the forge hammer from the balling furnace into the shape and form required for the use intended in one heat, after which it is again heated to a welding heat in the furnace, and passed under the rollers, for any of the purposes herein-before described, which makes good iron without the usual process of refining with charcoal or coke. Pig and sow metal, commonly called
25 cast iron, and old cast iron, is to be sunk down in the finery into half blooms, and shingled under the forge hammer to the intended size of the use in one heat, instead of being drawn into anchonies; after which it undergoes a like process by welding under the rollers, whereby it is perfectly refined, and at the same time reduced, either by the said rollers or, being passed from them
30 through the cutters, to the uses required; or both of these kinds of iron may be prepared for the furnace by methods now practiced, and then shingled and welded to uses in one heat under the rollers, in the method herein-above specified with respect to scrap iron. All which iron, prepared, welded, worked, and reduced into use, by the methods and processes before set forth, will be
35 equal, if not superior, in quality to iron made by the ordinary process or methods before in practice. And the nature of my said Invention, so performed by the said methods and processes herein-above set forth, especially consists in taking a welding heat on several pieces or parcels and faggots of iron in the same furnace at one time, and drawing the same into uses, either

A.D. 1783.—Nº 1351.

Cort's Method of Preparing, Welding, and Working Iron.

singly, or by welding and shutting two or more together by the operations described, either under the forge hammer or through the mill rollers, as may best suit the use intended. And these operations may be diversified so as to suit all purposes whatever to which forged or welded iron is or can be applied, as well as those herein particularized by way of example.

 In witness whereof, I, the said Henry Cort, have hereunto set my hand and seal, the Sixteenth day of May, in the year of our Lord One thousand seven hundred and eighty-three, and in the twenty-third year of the reign of our said Sovereign Lord George the Third, by the grace of God of Great Britain, France, and Ireland King, Defender of the Faith, &c.

HEN. (L.S.) CORT.

 AND BE IT REMEMBRED, that on the same Sixteenth day of May, in the year above mentioned, the aforesaid Henry Cort came before our said Lord the King in His Chancery, and acknowledged the Specification aforesaid, and all and every thing therein contained and specified, in form above written. And also the Specification aforesaid was stampt according to the tenor of the Statute made in the sixth year of the reign of their late Majestys King William and Queen Mary of England, and so forth.

 Inrolled the Sixteenth day of the same May, in the year above mentioned.

JOHN WILMOT.

LONDON:
Printed by GEORGE EDWARD EYRE and WILLIAM SPOTTISWOODE,
Printers to the Queen's most Excellent Majesty. 1856.

A.D. 1784 . . . . N° 1420.

Manufacture of Iron.

CORT'S SPECIFICATION.

TO ALL TO WHOM THESE PRESENTS SHALL COME, I, HENRY CORT, of Funtly Iron Mills, in the Parish of Titchfield, in the County of Southampton, the Grantee in the Letters Patent herein-after mentioned, do send greeting,
5 WHEREAS our Sovereign Lord King George the Third, by His Royal Letters Patent under the Great Seal of Great Britain, bearing date at Westminster, the Thirteenth day of February, in the twenty-fourth year of His said Majesty's reign, upon my Petition (whereby I had humbly represented to His said Majesty that I had, after great study, labour, and expence, and
10 in the prosecution of various discoveries, and by contriving and trying divers experiments, invented and brought to perfection " A NEW MODE AND ART OF SHINGLING, WELDING, AND MANUFACTURING IRON AND STEEL INTO BARRS, PLATES, RODS, AND OTHERWISE, OF PURER QUALITY, IN LARGE QUANTITIES, BY A MORE EFFECTUAL APPLICATION OF FIRES AND MACHINERY, AND WITH GREATER YIELD, THAN
15 ANY METHOD BEFORE ATTAINED OR PUT IN PRACTICE)," and His said Majesty, being willing to give such encouragement as in the said Letters Patent is mentioned, was graciously pleased to condescend to my request, and did therefore make known that His said Majesty, of his especial grace, certain knowledge, and meer motion, did give and grant unto me, the said Henry Cort, my executors,
20 administrators, and assigns, His said Majesty's especial licence, full power, sole

99

A.D. 1784.—N° 1420.

Cort's Manufacture of Iron and Steel into Bars, Plates, Rods, &c.

priviledge and authority, to use, exercise, and vend my said Invention, and touching and concerning the same, as in the said Letters Patent is in that behalf particularly mentioned and expressed, to have, hold, exercise, and enjoy the said licence, powers, priviledges, and advantages therein granted unto me, the said Henry Cort, my executors, administrators, and assigns, for and 5 during and unto the full end and term of fourteen years from the date of the said Letters Patent next and immediately ensuing, and fully to be compleat and ended according to the Statute in such case made and provided, to make, use, exercise, and vend my said Invention within that part of the Kingdom of Great Britain called England, the Dominion of Wales, and 10 Town of Berwick-upon-Tweed, and also in all the Colonies and Plantations abroad, conformable to His said Majesty's gracious pleasure and command, and to the several provisoes, declarations, restrictions, grants, and conditions in the said Letters Patent expressed and contained, as in and by the said Letters Patent under the Great Seal of Great Britain, dated as aforesaid, 15 relation being thereunto had, may more fully appear.

NOW KNOW YE, that I, the said Henry Cort, in compliance with the said Letters Patent, and in performance of a condition or proviso therein contained for that purpose, have described and ascertained, and by these Letters Patent do describe and ascertain, the nature of my said Invention, and the manner of 20 performing the same as followeth, that is to say:—

For the preparing, manufacturing, and working of iron from the ore, as well as from sow and pig metal, and also from every other sort of cast iron (together with or without scull and cinder iron and wrought-iron straps), I make use of a reverberatory or cur furnace or furnaces of dimensions suited 25 to the quantity of work required to be done, the bottoms of which are laid hollow or dished out, so as to contain the metal when in a fluid state. My furnace for the first part of the process being got up to a proper degree of heat by raw pit coals, or other fuel, the fluid metal is conveyed into the air furnace by means of ladles or otherwise. When this air furnace is charged with sow 30 and pig metal, or any other sort of cast iron, the door or doors of the furnace should be closed till the metal is sufficiently fused, and when the workman discovers (through a hole which he opens occasionally) that the heat of the furnace had made a sufficient impression upon the metal, he opens a small aperture or apertures, which I find is convenient to have provided in the 35 bottom of the doors (but which is or are closely shut, as well as the doors, at the first charge of the furnace with cold cast metal); and then the whole is worked and moved about through those apertures by means of iron bars and other instruments fitly shaped, and that operation is continued in such manner as

Cort's Manufacture of Iron and Steel into Bars, Plates, Rods, &c.

may be requisite during the remainder of the process. After the metal has been some time in a disolved state, an ebullition, effervescence, or such like intestine motion takes place, during the continuance of which a blueish flame, or vapour is emitted; and during the remainder of the process the operation
5 is continued (as occasion may require) of raking, separating, stiring, and spreading the whole about in the furnace till it looses its fusibility, and is flourished or brought into nature; to produce which effect, the operations subsequent to the fused state are the same, whether the fusion be made in the air furnace, or the metal be conveyed into it in a fused state, as first mentioned. As
10 soon as the iron is sufficiently in nature, it is to be collected together in lumps, called loops, of sizes suited to the intended uses, and so drawn out of the door or doors of the furnace, when all the small pieces that may happen to remain are also cleared away. It has been found by me to be a good method of using such small pieces last mentioned, and also scull or tinder iron, first broken into
15 smal pieces; and also all sorts of parings of iron plate, or other thin iron, and nut or bushel iron, commonly called wrought straps, to throw them into the furnace in various proportions during the operation of bringing the fused metal into nature, and before it is collected into loops, and as the whole charge of the furnace is raked and stirred about, these straps become lapped up in
20 the loops after the fused metal is flourished and got into nature. And the whole of the above part of my method and process of preparing, manufacturing, and working of iron is substituted, instead of the use of that finery, and is my Invention, and was never before used or put in practice by any other person or persons. The iron so prepared and made may be afterwards stamped
25 into plates, and piled or broke and worked in an air furnace, either by means of pots, or by piling such peices in any of the methods ever used in the manufactory of iron from coke fineries without pots; but the method and process invented and brought to perfection by me, is to continue the loops in the same furnace, or to put them into another air furnace or furnaces, and to heat
30 them to a white or welding heat, and then to shingle them under a forge-hammer, or by other machinery, into half blooms, slabe, or other forms, and these may be heated in the chaffery according to the old practice; but my new Invention is to put them again into the same or another air furnace or other furnaces, from whom I take the half blooms, and draw them under the
35 forge-hammer, or otherwise, as last aforesaid, into anthonies, barrs, half flatts, small squared tilted rods for wire, or such uses as may be required; and the slabe having been shingled in the foregoing part of the process to the sizes of the grooves in my rollers, through which they are intended to be passed, are worked by me through the grooved rollers, in the manner which I use bar or

A.D. 1784.—N° 1420.

Cort's Manufacture of Iron and Steel into Bars, Plates, Rods, &c.

wrought iron fagotted and heated to a welding heat for that purpose. Which manner of working any sort of iron in a white or welding heat through grooved rollers, is entirely my own Invention. Iron and also steel so prepared, made, wrought, and manufactured, by such effectual application of fire and machinery, will be discharged of the impurities and foreign matter which adhers to them when manufactured in the methods commonly practiced. The steel is of an excellent quality, and the iron will be found to be good tough iron in barrs and uses, whether large or small, and in all sorts of merchant iron, whether it be made from mettal of a red short or cold short nature; and blistered steel, whether made from iron prepared according to the above process, or from any other iron when faggoted together, heated to a white or welding heat, rolled in that heat through grooved rollers, according to the method invented by me, and slit through the common cutters, is equal to steel manufactured by forge and tilt hammers. The whole of which discovery and attainment are produced by a more effectual application of fire and machinery, as described by me, than was before known of or used by others, and are entirely new and contrary to all received opinions amongst persons conversant in the manufactory of iron; and the whole of my method may be compleated without the necessity of using finery, charcoal, cokes, chaffery, or hollow fire, and without requireing any blast by bellows or cillinders, or otherwise, or the use of fluxes in any part of the process. The whole opperation to be performed with one or more furnaces, according to the quantities or dispatch required.

In witness whereof, I, the said Henry Cort, have hereunto set my hand and seal, Twelfth day of June, in the year of our Lord One thousand seven hundred and eighty-four.

HEN^Y (L.S.) CORT.

AND BE IT REMEMBRED, that on the Twelfth day of June, in the year above mentioned, the aforesaid Henry Cort came before our said Lord the King in His Chancery, and acknowledged the Specification aforesaid, and all and every thing therein contained and specified, in form above written. And also the Specification aforesaid was stampt according to the tenor of the Statute made in the sixth year of the reign of their late Majesties King William and Queen Mary of England, and so forth.

Inrolled the Twelfth day of June, in the year above written.

LONDON:
Printed by GEORGE EDWARD EYRE and WILLIAM SPOTTISWOODE,
Printers to the Queen's most Excellent Majesty. 1856.

INDEX

Anchors
    faggoting in manufacture of, patent   97
    made from Cort's iron tested by direct pull   43
ATTWICK, WILLIAM, of Gosport
    Cort agent to   19
    iron supplied to Portsmouth Dockyard by   20
    uncle of Cort's first wife   19

BLACK, Dr JOSEPH
    description of Cort's process   44
Black Country
    iron castings trade not immediately displaced by puddling   67
Bloomfield Ironworks
    Joseph Hall introduces wet puddling   76
Bursledon furnace
    blown out in 1772   26

Chains
    made from iron ballast by Cort's process   43, 45
charcoal
    copses for making, in Hampshire   25
    transport by water from Beaulieu   25
coal
    use in pot refining process   3
    use in reverberatory furnace   5, 8, 9
Coalbrookdale group
    failure of Onions's process at works of   13, 47
    refusal to pay Cort royalties   50
Cockshutt's process
    two-stage refining, using pit coal fining followed by charcoal melting   11
coming to nature
    Schubert's explanation   15
CORT, CONINGSBY
    failure in petition for Government award   68–71
CORT, HENRY
    bankruptcy

Cort, Henry *(continued)*
        never discharged   63
        proceedings in   63
        valuation of stock and property at   61, 62
    birth probably illegitimate   xiv
    children of   18
    correspondence with Richard Crawshay   53, 54, 56
    death   63
    debts paid off by Samuel Jellicoe   63
    marriages   16, 17
    mast hoops
        contract   28
        contract, loss on   29
        rolling in manufacture of   29, 30
    as Navy Agent   19
    pension from William Pitt   64, 65
    royalty offer from Richard Crawshay   56
CORT, RICHARD
    Civil List pension, 1855   66
Cort family
    Palmerston recommends pensions   79
Cort memorial plaque
    to be erected in 1983   79
Cort memorial tablet
    unveiled Hampstead, 1905   79, *frontispiece*
Cort's patents
    denigrated by Samuel Homfray at Parliamentary enquiry   68
    estimated value less than development costs   67
    specifications   91, 99
    valuation at bankruptcy   62, 67
Cort's tombstone
    lettering restored 1966   79, *plate 5*
Cort's unmarried daughters
    pensions 1816, increased 1856 and 1859   66
Cort's wrought iron
    compared with Swedish iron   40
    faggoting of   41
    tests on   41, 43
    wide economic influence in nineteenth century   88
Cranage process
    pit coal used in reverberatory furnace   8
    trial of, at Coalbrookdale   9
CRAWSHAY, RICHARD
    correspondence with Henry Cort   53, 54, 56
    correspondence with Adam Jellicoe   53
    new furnace at Cyfarthfa   73

Index

offers Cort 5s a ton royalty   56
visits Fontley   53
Cumberland
    haematite mines of Bigrigg and Cleator Moor   2
    Whitehaven coalfield   2, 4, 5
Cyfarthfa forge
    pot process employed   7
Cyfarthfa Ironworks
    largest in Europe   73
    production figures   73
    trials of Cort's process at   53, 54, 55

DARBY, ABRAHAM
    use of coke in blast furnace   xiii
dockyards, Royal
    iron requirements   19

Elizabethan period
    use of water-powered slitting mill in Thames Valley   32

Faggoting
    in anchor manufacture, patent   97
    in hoop manufacture   41
Fareham ironworks
    Fontley (Titchfield)   22, 24
    Gosport, smith's shops   22, 26
    Titchfield (Fontley)   22, 24
Fontley (Titchfield)   22, 24
    Cort leases   27
    earlier lessees   24
    relation to Bursledon, Gosport, etc.   *plate 4*
    waterwheel at   27, 28
fining processes
    Cockshutt's   11
    Cranage's   8, 9
    Onions's   13, 14
    Roebuck's   10
    Wood family's pot processes   1, 2, 3, 4, 6, 7
    Wright and Jesson's   11, 12
forges
    capacity in eighteenth century   24
Funtley, Fontley—*see legend plate 4*

Gringoe family
    lessees of Bursledon furnace, 'Titchfield Hammer'
    and quay at Fawley   24, 25

grooved rolls
    Cort's use in finishing and for producing rounds and other shapes   31, 35
    Harvey's patent for bolt finishing   34
    mill with   *plate 8*
    Payne's patent, proposing use of windmill   34
    Polhem's reference to   34
    Purnell's specification   35, *plate 3*
    for rolling lead   34

HALL, JOSEPH
    introduces wet puddling at Bloomfield ironworks   76
HAMMOND, Sir ALEXANDER,
    Comptroller of the Navy, authorises pension for Mrs Cort   66
HARTLEY, DAVID
    description of Cort's process   45
HARVEY, THOMAS
    patent for finishing in grooved-roll mill   34
hollow fire
    explanation of term   10, 11
HOMFRAY, JEREMIAH
    operates Cort's process at Cyfarthfa   55
HOMFRAY, SAMUEL
    denigrates Cort's process   68
    fails to operate Cort's process on Plymouth furnace pig   54
    a litigious person   70

Iron ballast
    converted by Cort's process into mooring chains, etc.   43, 45
iron ore, Cumberland   2
ironstone, Hampshire
    outcrops and mining   25
iron, wrought
    Navy requirements   19, 20, 21
    Swedish imports   19–21, 29, 40

JELLICOE, ADAM
    correspondence with Richard Crawshay   53
    Cort's debts to   60
    death brings about Cort's bankruptcy   57
    use of naval balances to finance Cort   60
JELLICOE, SAMUEL
    correspondence with Richard Crawshay   53
    pays off debts of Henry Cort and Adam Jellicoe   63

Ketley ironworks
    Cort's process tried at   48, 49

Index

Low Moor ironworks
    coke refinery   *plate 6*
    puddling introduced   75

Merthyr Tydfil
    iron production increases in nineteenth century   74
MONTAGU, DUKE of
    leases Sowley ironworks to various operators   25

Navy, iron requirements   19, 20, 21
Navy Board
    evades responsibility of determining whether Cort's patents were being infringed   64

Onions's process
    coal refining, unsuccessful   13
    Percy's misunderstanding of   14
Oregrund iron   *see* Swedish iron

PALMERSTON, LORD
    recommends pensions for Cort family   79
Parliamentary enquiry 1812
    into case for government award to Cort family   67–71
patents, Cort's specifications   91, 99
PAYNE, JOHN
    patent for grooved-roll mill to be operated by windmill   34
pig-iron
    manufacture   xi
    refining   *see* fining processes
PITT, WILLIAM
    obtains pension for Cort   64, 65
Plymouth furnace
    connected with Cyfarthfa   8
Plymouth pig-iron
    Cort makes good wrought iron, Samuel Homfray fails   54
POLHEM, CHRISTOPHER
    inventor of back-up rolls   34
    reference to grooved rolls   34
puddling furnace   *plate 7*
puddling process
    change from dry to wet puddling   76
    Cort's, descriptions   37, 38, 44, 45
PURNELL, JOHN
    grooved-roll mill, specification   35, *plate 3*

REYNOLDS, WILLIAM
    trials on Cort's wrought iron   48, 49
    trials on Onions's process   47, 48
Roebuck's process
    two-stage refining with pit coal   10
rolling
    nineteenth century refinements   77
rolls
    grooved   *see* grooved rolls
    reversing   *plate 1*

SCHUBERT, H. R.
    explanation of term 'come to nature'   15
scrap, iron
    worked up by pot refining and the chafery   7
sheet rolling   32, 33
SHEFFIELD, LORD
    advantage of Cort's process to the iron trade of Great Britain   46
slitting mill   *plate 2*
    for nail rod   32
    for plate   32, 33
    for silver strip   32
Sowley forge and furnace   24, 25
    lease of, by Duke of Montagu   25
SOUTHAMPTON, EARLS of
    encourage ironmaking at Titchfield and Sowley   22
Swedish iron
    compared with Cort's   40
    used in dockyards, prices   21

Testing of Cort's iron by direct pull of anchors   43
three-high mill
    developed in Staffordshire   77
TROTTER, ALEXANDER
    Paymaster of the Navy   58
    use of Naval balances for his own purposes   58
    writ against Adam Jellicoe leads to Cort's bankruptcy   60

Wire drawing, hand and mechanised   31, 32
Wood family, pot refining process   1, 2, 3, 4, 6, 7
Wright and Jesson's process, four-stage refining   11, 12
wrought iron, wide economic influence of Cort's process for   88